The Church of St Mark, at Venice. Page 35.

BOSTON
D. LOTHROP & CO.
38 & 40 CORNHILL.

Building Stones.

BY

MRS. JULIA P. BALLARD.

"Behold, I lay in Zion, for a foundation, a stone."
ISA. xxviii. 16.

"Ye, also, as lively stones, are built up a spiritual house."
1 PET. ii. 5.

"In whom ye also are builded together for an habitation of God."
EPH. ii. 22.

"Other foundation can no man lay than that is laid."
1 COR. iii. 11.

Boston:
Published by D. Lothrop & Co.
Dover, N. H.: G. T. Day & Co.
1871.

Stereotyped at the Boston Stereotype Foundry,
No. 19 Spring Lane.

CONTENTS.

BUILDING STONES.

CHAPTER I.

THE FOUNDATION STONE.

THE silver reel of time had rolled off the months of another year, and snapped on the New Year's morning of 1870.

The day after New Year's was the Sabbath. In a pleasant room, gathered about a cheerful, open fire, sat Mrs. Ewing and her three children, Ernest, Olive, and Mary. It was a stormy day, and they had just come in with wind-painted cheeks, and hair blown about and powdered with flying snow-flakes; and, after removing their cloaks and clouds, and laying down their Sunday school books, had seated themselves by their mother before opening them.

"You have not forgotten our 'Golden Hour,'" said Olive, as their mother closed the book she was reading, and looked at her watch.

"We don't want to begin our books until we have that," said Mary. "We had a splendid time, in Esther, last Sabbath. I wish we had another place as good as that. Where shall we read to-day?"

"I think we might find a great many as good," replied Mrs. Ewing. "But suppose we do not, for this hour, read any whole or particular chapter. I would like to ask you some questions to-day on a verse or two in which I have become especially interested while you were at school. But how about Ernest joining us?" she asked, looking pleasantly at him in a questioning way.

The Golden Hour. Page 7.

Ernest was just that moment busy with "One Day's Weaving," glancing here and there through the book he had been anxious to get for two or three weeks; but he heard the question, and closed it quickly as his mother spoke.

What the children meant by the "Golden Hour" was the first hour after they returned from Sabbath school, every Sunday, which their mother spent with them in reading some part of the Bible, which either she or they selected, and which they talked over freely, and greatly enjoyed. Mrs. Ewing was herself a dear lover of the Bible. It was to her a solid delight to explore its hidden treasures with her children, and lead them, with pleasure and without compulsion, into the secret places of truth.

Sometimes, as Olive said, "the hour ran over;" but, as they did not usually find it out until the tea-bell rang, it did not make much difference.

"What sort of questions were you going to ask us, Mother?" said Olive. "Where shall we turn in our Bibles?"

"Before you open them at all," replied Mrs. Ewing, "I will ask either of you to mention some title by which Christ is called in the Bible. I have been thinking a good deal about one of these to-day."

"He is called a Lion," said Ernest.

"And a Lamb," said Olive.

"And a Rose," added Mary.

"And a Vine, isn't He?" asked Olive.

"And a Lily," said Mary.

"And a Bridegroom," said Ernest.

"And I know another," said Mary. "He is called a Star, — the Star of Bethlehem."

"Yes. You have thought of more than I supposed you would," said Mrs. Ewing; "but you have not yet named the one I was thinking of. Now we will open our Bibles. And, Ernest, will you read, when we have all found the sixteenth verse of the twenty-eighth chapter of Isaiah?"

Ernest soon found the place, while Olive and Mary turned over the leaves of their Bible, and were ready to follow him as he read: —

"Therefore, thus saith the Lord God, Behold I lay in Zion, for a foundation, a stone, a tried stone, a precious corner stone, a sure foundation; he that believeth shall not make haste."

"I don't think I should like the 'Stone' title so well as the 'Rose,'" said Olive.

"Or the 'Lamb,'" said Mary.

"Or the 'Lion,'" said Ernest.

"Or the 'Lily,'" added Olive.

"I am glad to have you speak freely, just as it seems to you," said Mrs. Ewing. "But let us see; why is He called a Lily?"

"I suppose, because He has the purity of the Lily," said Mary.

"And a Rose," said Olive, without waiting to be asked, "because He has the fairness and the sweetness of a Rose."

"And a Lion," added Ernest, "because of the greatness of His strength; and a Lamb, because He was meek and gentle."

"Yes; you are all right. Because of His purity in the midst of a perverse generation, He might very beautifully be compared to a 'Lily among thorns.' As a Lion, He will be the destroyer of His enemies. As a Lamb, He was 'led to the slaughter,' and 'opened not His mouth.'"

"But, on thinking them all over, I believe we shall find a new and very great pleasure in looking at Him very carefully under the name of a 'STONE.' He has the beauty of flowers, but He has more; He has the meekness of the Lamb, but not its frailty; the strength of a Lion, but not his cruelty."

"A stone or rock indicates strength, as does the Lion," said Ernest; "but I like the Stone better."

"God calls Him here a 'Foundation Stone,'—'a Sure Foundation.' Now, what is to be built upon this stone?"

"His church in the world, I suppose, is it not?" asked Ernest.

"Yes. And what is one thing necessary for a foundation stone?" asked Mrs. Ewing.

"Strength," said Ernest.

"Yes. And would you not think, in laying the foundation of a great and important building, it would require great wisdom to select a stone that would be suitable for the place,—one that would bear whatever strain might come upon it, however high the building might rise, or whatever pressure might be brought to bear upon it? Now, who selected this stone? Let us turn to this verse in the 28th of Isaiah,—for it is a very wonderful verse,—and see. 'Thus saith the Lord God, Behold I lay in Zion.' Who, then, chose the stone?"

"God," replied Ernest.

"Yes. 'Thus saith the Lord God.' We can have no higher authority than this. And what does He say, first, of the stone?"

"He says, 'I will lay in Zion,'" said Olive.

"Then, if God, who is infinite in wisdom, chose the foundation stone, and laid it Himself, may we not be very sure it will be the best that could possibly have been selected? It may well be called a sure foundation. Upon Christ His people may securely build. He will never prove a false foundation. He is also a corner stone. The corner stone unites the different parts of a building together. Christ, as a chief corner stone, connects and establishes his church. On this foundation, both Jews and Gentiles build; from this corner stone they are built up, one church complete in Him. Is there any other safe foundation for the church to build upon, Ernest? You may turn to 1 Cor. 3 : 11., and read an answer to my question."

Ernest found the place quickly, and read, "Other foundation can no man lay than that is laid, which is Jesus Christ."

"How plain that is!" said Mary.

"I should think," said Ernest, "that would be a lesson to those who think Peter was a rock strong enough to build upon."

"Other foundation can no man lay," repeated

Mrs. Ewing. "Would that these words could be engraved upon your hearts, as upon a rock, with a diamond's point. Would that they could be spoken with force in the ear of every sinner who is longing to be saved, and casting about for some secure place upon which to build his hopes of eternal life. Does Christ Himself ever allude to this title of the Stone, as applied to Him by God, Ernest?"

"I am not sure."

"Turn to Matt. 21 : 42."

Ernest read, "Jesus saith unto them, Did ye never read in the Scriptures, The stone which the builders rejected, the same is become the head of the corner? This is the Lord's doings, and it is marvelous in our eyes."

"What did He mean by that?" asked Olive.

"He was talking to the priests and elders in the temple at Jerusalem. They came up to him as He was teaching, and said, 'By what authority doest thou these things, and who gave thee this authority?' Then He repeated to them the parable of the householder, who planted a vineyard, and digged a wine-press in it, and built a tower, and let it out to husbandmen, and went into a far country, and afterward sent his servants, and then his son, to gather for him the fruits of the vineyard. But they beat, stoned, and killed the servants, and killed the son, also. He left them to make the application, adding, however, these striking words which you have just read,—'The stone which the builders rejected, the same is become the head of the corner.'"

"Where were these words spoken before?" asked Olive. "He said, 'Did ye never read in the Scriptures?'"

"He was quoting from David, in Psalm 118. You may read the 22d and 23d verses of that Psalm," said Mrs. Ewing.

Olive turned to the place and read, "The stone which the builders refused is become the head stone of the corner. This is the Lord's doing; it is marvelous in our eyes."

"I think I know what it means," said Ernest. "That God's plan of salvation for the world was through Christ, this Stone; and the Jews, who looked so long for Him, rejected Him when He came; but the work went on all the same without them, and Christ was honored as the true foundation of the church."

"You are right. Turn to Acts 4 : 11, and see what Peter there plainly tells the Jews about their treatment of Christ."

Ernest read: "This is the stone which was set at naught of you builders, which is become the head of the corner; neither is there salvation in any other; for there is none other name under heaven, given among men, whereby we must be saved."

"That is the same," said Olive, "as, 'Other foundation can no man lay.'"

"Yes. Is it not strange that any person should take Peter for a foundation, — the very man who was so earnest in setting Christ before men as the only foundation?" said Mrs. Ewing.

"I never thought of that before," said Ernest.

"It seems so strange," said Olive, "that when the Jews had so long looked forward to the coming of a Saviour, they should fail to receive Him when he really came."

"It was hard for them to receive one coming as a babe in a manger, when they were looking for a great king. In Mark 9 : 11, His disciples seem to have been questioning in their own minds about Christ, and thinking of the prophecies with regard to Him; and they asked Him this question: 'Why say the scribes that Elias must first come?'"

"Whom did they mean by 'the scribes'?" asked Mary.

"They were learned men among the Jews, who wrote the law in Hebrew, and taught it. They were very particular in preserving every letter of the law carefully. It was their great care to know what was of real value, and keep it before the people. Ezra was a noted scribe.

The King of Persia, Artaxerxes, speaks of him, in Ezra 7 : 12, as 'a scribe of the law of the God of heaven.' The disciples meant that it had been foretold that Elias should come before their Saviour came, did they not?" asked Olive.

"Just that."

"Where was that prophecy?" asked Ernest.

"You may find it in the very last chapter of the Old Testament. It was the last recorded promise in their Scriptures which was given to the Jews before Christ came," said Mrs. Ewing.

"No wonder they looked for its fulfillment, then," said Ernest. "I have found the place in Malachi 4 : 5, 6." He read, —

"Behold, I will send you Elijah the prophet before the coming of the great and dreadful day of the Lord. And he shall turn the heart of the fathers to the children, and the heart of the children to their fathers, lest I come and smite the earth with a curse."

"Those are the very last words of the Old Testament," said Ernest.

"Yes," replied Mrs. Ewing, "and that was what the disciples meant by their question, 'Why say the scribes (that is, the writers of the Bible) that Elias must first come?' Jesus told them, very plainly, why they had said so. In the next verse (Mark 9 : 12, 13) he answers them: 'Elias verily cometh first, and restoreth all things; and how it is written of the Son of Man, that he must suffer many things, and be set at naught. But I say unto you that Elias is indeed come, and they have done unto him whatsoever they listed, as it is written of him.' In Matt. 11 : 14, He told them plainly that John the Baptist was Elias. In Matt. 11 : 10, Jesus tells them that Isaiah meant John in his prophecy, which he then quoted to them from Isa. 40 : 3. 'For this is he of whom it is written, Behold, I send my messenger before thy face, which shall prepare thy way before thee.' Isaiah says, 'The voice of him that crieth in the wilderness, Prepare ye the way of the Lord, make straight in the desert a highway for our God;' and in Malachi 3 : 1, almost the exact words which Christ quoted are found: 'Behold, I will send my messenger, and he shall prepare the way before me;' and it is further added in Malachi, in the same place, — 'And the Lord, whom ye seek, shall suddenly come to His temple, even the messenger of the covenant whom ye delight in.'

"They could neither see Elias in John, nor their Saviour in Jesus. How touching are the words he uttered, in Matt. 11 : 14: 'And if ye will receive it, this is Elias which was for to come.' 'If ye will receive it'! This was not only said to the Jews in Christ's day, but He says it to all to-day who doubt and fear to believe in Him. 'If ye will receive it, I am your Saviour.'"

"I suppose," said Olive, "He meant Elias for one of those who came to the vineyard looking for fruit, and was killed; and He Himself was 'the Son,' was he not?"

"Yes. And well might the Lord of the vineyard have said, — 'They will reverence my Son.'"

"But they did not," said Olive.

"No; even He was 'set at naught' of the builders. The Jews, to this day, look for the coming of Elijah, as well as for a Messiah. It is their custom, at some of their feasts, to open the door and call for Elijah."

"I remember," said Ernest, "that converted Jew who was here, spoke of this in one of his lectures. He described a Jewish family eating the Passover, and at a certain time during the meal one of their number would go to the open door and call for Elijah."

"It shows how strong their faith was in their law; so far as they had accepted the word of God, so far they believed it to the very letter. It makes it all the more sad that for so many the 'vail is still over their faces' with regard to Christ. The Jews were looking for their Saviour, too, at the very time

He came, and were expecting Elijah so confidently that a common expression among them was, 'Wait till Elijah comes.' To those who did not receive Christ as their foundation stone, he is also called a Stone; but something very different from a 'corner stone,'" continued Mrs. Ewing.

"What kind of a stone?" asked Mary.

"You will find an answer in Isaiah 8: 13, 14. I will read to you. 'Sanctify the Lord of Hosts Himself; and let Him be your fear, and let Him be your dread. And He shall be for a sanctuary; but for a stone of stumbling, and for a rock of offense to both the houses of Israel; for a gin and for a snare to the inhabitants of Jerusalem. And many among them shall stumble and fall, and be broken, and be snared, and be taken.' Peter quotes this verse from Isaiah, about the foundation stone, and joins the two, — clearly showing to whom Christ is a foundation stone, and to whom a 'stone of stumbling.' Will you read it, Ernest, in 1 Pet. 2: 6–8?"

Ernest read:—

"Wherefore also it is contained in Scripture, Behold, I lay in Zion a chief corner stone, elect, precious; and he that believeth on Him shall not be confounded. Unto you, therefore, which believe, He is precious; but unto *them which be disobedient, the stone* which the builders disallowed, the same is made the head of the corner; and a stone of stumbling, and a rock of offense, even to them which stumble at the word."

Here the tea-bell showed them that the hour had "run over" for this afternoon.

"I would not have thought we could be so much interested in just one verse," said Ernest.

"And no story," added Olive.

"Except the 'old, old story,'" said Mrs. Ewing, "which is, after all, the sweetest. But we have not yet got through with this one-verse story."

CHAPTER II.

THE TRIED STONE.

THE next "Golden Hour" came almost before the children thought of it. No sooner were they home from Sabbath school than their Bibles were all opened to Isaiah.

"To-day," said Mrs. Ewing, "we will talk about this stone as a 'tried' stone. But, first, I want to see if you remember what sort of a stone Christ is to those who do not believe in Him; Olive?"

"A stumbling stone."

"You see from this, that it is of the greatest moment in which of these ways we shall take

Christ. By our own choice we shall either be safely built upon Him, or, by unbelief, find Him a rock of offence over which we shall stumble and perish. Ernest, will you turn to the ninth chapter of Romans, and read the last three verses?"

Ernest read: —

"But Israel, which followed after the law of righteousness, hath not attained to the law of righteousness. Wherefore? Because they sought it not by faith, but as it were by the works of the law. For they stumbled at that stumbling stone; as it is written, Behold, I lay in Zion a stumbling stone and rock of offense; and whosoever believeth on Him shall not be ashamed."

"Those are solemn words," said Mrs. Ewing, "but they close with a precious promise: 'Whosoever believeth on Him shall not be ashamed.'"

"I've got an answer for Jasper Hall," said Ernest, looking up quickly.

"What is that?" asked his mother.

"He says, 'it does not make any difference what a boy believes, if he is only sincere.' He was arguing that, only yesterday, with Fred Green and me, and I could not answer him very well; though I felt sure he was not right."

"What did you say?" asked Mrs. Ewing.

"A good many things. At last I said I thought I could prove it in one way. I said, 'If I see some scarlet berries, and I believe they are winter-green berries, and feel perfectly sure of it, and eat them, if they are not winter-green berries, but a poison berry, will I be safe because I believed they were good to eat?"

"I should think he could see that," said Olive.

"But he said 'he wasn't talking about such things, but about what we believe as a guide to doing what we call right and wrong.' Now I felt pretty sure that my answer puzzled him, though he wouldn't admit it. But here he could not say anything."

"What do you mean?" asked Olive.

"I mean it did not save the Jew from stumbling over Christ into ruin because he believed that He was an impostor, and not the true foundation stone on which he was safest to build. And neither was it because they believed this, that those who did build on Christ were saved merely; but *because they believed what was true.*"

"You have put that very well, Ernest," said his mother. "I hope you will never forget your own argument. Find what is truth, and then believe it with your whole heart. This brings us to Christ as a 'tried' stone."

"I never knew what a 'tried' stone meant," said Mary.

"There is a great difference in stones. Some are weak, some strong, some hard like flint, some soft and crumbling like sandstone. There is, of course, a great difference in their power to bear weights put upon them, and in their power of resistance to the various influences which might tend to weaken or destroy them. It is very easy to be deceived in judging stones. Some may look much harder than they really are. In this verse we have in one word what is necessary, first of all, in choosing a stone to be used as a foundation or corner stone."

"Tried," said Olive.

"That is it; God here says of Christ that He is a tried stone."

"In what way are building stones tried?" asked Ernest.

"In different ways. There are some things besides strength to be thought of. The atmosphere affects stones differently. Sea air causes some kinds of stones to disintegrate, or fall apart. Heat and frost affect different kinds of stones differently. Some stones are hardened by heat, while others are cracked and made useless by it. All these things are taken into consideration by those who select stones for important places in a building. The weight which different kinds of stones will support is tried by pressure; immense weights are brought to bear upon them; and tables have been given showing the different degrees of pressure different kinds of stones will bear."

"What stone bears the greatest weight?" asked Ernest.

"Porphyry."

"What kind of stone is that?" asked Olive.

"A variety of marble. It is very compact, and is found in many different colors. It has been ascertained that one square foot of porphyry will safely bear the pressure of six hundred and forty thousand pounds."

"How is this known?" asked Ernest.

"The weight which a square foot will bear is found by getting the weight of the stone, and then subjecting it to the greatest possible pressure which this one square foot will bear. Then one eighth of the amount of pressure that will crush the stone may be considered a safe weight to be put upon it. The value of this method of testing stones is very great. From a want of this knowledge, and the consequent insertion of stones which will not support the amount of pressure required, buildings have fallen in a comparatively short time which otherwise might have stood for ages. Buildings, where the foundation stones are well chosen, have been known to resist the action of the air, and the pressure of the weight brought to bear upon them, to a degree almost beyond our belief. In the old French town of Angers, there is a building,— the 'Church of All Saints,'— the pillars of which, it is said, support on each superficial foot a pressure of eighty-six thousand pounds. The Bagreaux stone, in the four triangular piers of the great dome of the Pantheon, or St. Genevieve, at Paris, bears a weight of sixty thousand pounds. The dome is of solid stone, and is triple, having an inner, outer, and intermediate dome. This Pantheon rises one hundred and ninety feet from the ground."

"I should think they would need 'tried stones' in such a place," said Olive.

"Is a Pantheon the same as a church?" asked Alice.

"It is a church for gods, — a building in which the worship of all the gods was conducted. The word itself is from two Greek words, meaning 'all' and 'god.' The Pantheon at Rome is the most celebrated one in the world. It was built twenty-six years before Christ, and was taken six hundred years after Christ and used as a Christian church. It was one of the wonders of the world. Its dome is very grand, and its top is nearly one hundred and fifty feet from the ground. The Pantheon is a round building, four hundred and twenty-nine feet in circumference. It has a Corinthian portico, so beautiful in its construction as to have been pronounced 'more than faultless.' It is one hundred and ten feet in length, and forty-four feet deep, and is composed of sixteen columns of granite, each nearly fifty feet high and fifteen feet in circumference. These supported a roof of bronze. This building, after the passing away of nearly two thousand years, is still the least impaired of all the ancient buildings of that great city."

"I am glad it became a Christian temple," said Olive.

"It is wonderful that it should have remained until now so well preserved. But it was built for ages. Its solid masonry mocks the touch of Time; even its cornices of marble and its mosaic pavements are said to be so well kept as to give one now an excellent idea of its first wonderful beauty and magnificence."

"I wish you would tell me one thing," said Ernest. "It always puzzles me when I read or hear of Corinthian pillars, and Gothic arches, and Doric architecture. I wish I understood exactly what it means. I mean, how the different kinds were built, and whether they were united in the same buildings."

"We shall get a little way from our 'foundation stone,' perhaps," said Mrs. Ewing; "but as you have asked the question, I will try and explain it to you. It is an excellent idea, too, if it can be done without too much interruption, never to go on talking about what we do not understand."

"I think I know what the Gothic kind of building is," said Mary; "because my German box of blocks has a temple that Aunt Fanny told me was Gothic. I can put it together with pointed roof and turrets, and make a beautiful church."

"That is right; and if you will bring me your box of blocks, I will show you all, just here, the different kinds or orders of building."

Ernest smiled as Mary came tugging in her heavy box, but Mrs. Ewing took it from her, and, clearing the books from the table, proceeded with her building, much to Mary's delight.

"What if any one should come in now?" said she, laughing. "They would think we were all playing on the Sabbath."

"But if we are not," said Mrs. Ewing, pleasantly, "they would only be mistaken, and we should not be hurt. Of course there are a great many varieties of building," continued their mother. "They have, however, been reduced to a very few, as the roots of all the rest. One writer says that the two orders of architecture, the Doric and the Corinthian, are the roots of all European architecture, which is itself derived from Greece, through Rome, and perfected from the East. The Doric is the groundwork of all Roman architecture. The Corinthian embraces all Gothic, old English, German, French, and Tuscan styles.

"The Doric is so named from Doris, or the Dorians, in Greece. It is noted for its strength and simplicity. It was used almost universally among the Greeks. Take these square blocks. I lay down one, then another of the same shape upon it, only a size smaller, and a third still a size smaller upon that. This forms the base, or stylobate, as it is called. Upon the top of this base stands the column, gradually decreasing in size to the top; this is plain or fluted. Upon the columns rests the entablature, which is also divided into three parts. I lay this block upon the columns for the roof, being the first part, or architrave, above it, so; another, the frieze, of the same hight; and, lastly, the cornice. This plain, straight, square work, capable of endless variations in detail, is the Doric. The Gothic is the pointed work, — sharply-pointed arches and vaulted roof. This originated with the North people, or Goths, of Norway and Sweden. The Ionic is Grecian, and has a lighter column, more slender and graceful than the Doric. The Corinthian is also Grecian, and is still more delicate than the Ionic. The capital, or upper part of the column, is usually carved in leaves of the acanthus and olive. This gives it a light and airy look, with its clambering leaves running up its windings, nestling in its angles and hollows, and clasping its stony shafts with stony blossoms, which natural leaves would have delighted to clasp."

"I should like the Corinthian best," said Olive.

"And this is my Corinthian temple," said Mary, who had been taking out the straight, Doric pillars from her mother's work, and putting in some prettily carved ones in their place. "I do not know that I have made it very plain to you, but you will have some better idea of the principal kinds, perhaps; and when I can do so, I will show you a book of engravings representing these different styles of architecture in detail."

"That I should like," said Ernest; "and now are these kinds ever blended in one building?"

"Yes. And sometimes all the principal kinds meet in one, as in the church of St. Mark, at Venice. I have read a very beautiful and minute description of that recently. Its foundations were laid in 977, and it was consecrated in 1111. During the time of its building, over a hundred years, every vessel from the East was required to bring material, — marble and pillars, — for this building. Its form is that of the Greek cross, with the addition of porches. It is surmounted by five domes; the middle one is ninety feet in hight, and the four others are eighty. The interior is most elegantly finished, with columns, odd walls of precious marble, and floor of wonderfully beautiful Mosaic; — 'a multitude of pillars and white domes, . . . hollowed into five great vaulted porches, ceiled with fair mosaic, and beset with sculpture of alabaster, clear as amber and delicate as ivory; sculpture, fantastic and involved, of palm leaves, and lilies, and grapes, and pomegranates, and birds clinging and fluttering among the branches, all twined together into an endless network of buds and plumes. . . . And round the walls of the porches there are set pillars of variegated stones, jasper and porphyry, and deep green serpentine, spotted with flakes of snow; . . . the capitals rich with interwoven tracery, rooted knots of herbage, and drifting leaves of acanthus and vine, and mystical signs, all beginning and ending in the cross; and above them, in the broad archivolts, a continuous chain of language and of life-angels, and the signs of heaven, and the labors of men, each in its appointed season upon the earth; and above these, another range of glittering pinnacles, mixed with white arches, and edged with scarlet flowers.'"

"O, how beautiful!" said Olive.

"But stranger than its beauty, I thought," said Mrs. Ewing, "is the indifference with which those who live in its shadow come to regard it. The same writer says, 'You may walk from sunrise to sunset, to and fro, before the gateway of St. Mark's, and you will not see an eye lifted to it, nor a countenance brightened by it. Priest and layman, soldier and civilian, rich and poor, pass by it alike regardless. Round the whole square in front of the church there is almost a continuous line of cafés, where the idle Venetians of the middle class lounge, and read empty journals; . . . and in the recesses of its porches, all day long, knots of men of the lowest classes, unemployed and listless, lie basking in the sun like lizards; and unregarded children, — every heavy glance of their young eyes, full of desperation and stony depravity, and their throats hoarse with cursing, — gamble and fight and snarl and sleep, hour after hour. . . . And the images of Christ and his angels look down upon it continually.' "

"How strange that is!" said Ernest.

"Yes; and I have read it to you because it struck me so very forcibly as a picture of the way in which the world now regards this wonderful building of the church upon Christ, the great foundation stone. Men do not see it, or do not regard the beauty of either its wonderful foundation or the slowly rising structure upon it, struggling up through the centuries, in the midst of evil, and every hindering work. But it will yet be finished with completeness and marvelous beauty, and God shall crown it with joy."

"How did we get to the subject of palaces and temples?" asked Olive.

"I do not wonder you ask," said Mrs. Ewing, with a smile. "We were talking of tried stones, and I was telling you what weight certain stones bore in different buildings, when we turned from the test stones to the palaces themselves."

"You said the stone in the pillars of the Paris Pantheon bore up a weight of sixty thousand pounds?"

"Yes; and the piers under the dome of St. Paul's, in London, sustain thirty-nine thousand pounds weight, and those of St. Peter's even more. It is said that stones which have been exposed for nearly one thousand years in the Houses of Parliament and in Southwell Minster, still retain every mark of the tools with which they were cut; while the same kind of stone, in London, soon suffers from the sulphurous acids contained in the smoke of the city."

"I never thought how much the word 'tried' meant before," said Olive.

"Everything, in the way of beauty, duration, and safety. In our own country such experiments have been made, by scientific men, upon the marble used in the Washington Monument, as to cause them to prophesy that it will fall to pieces before its completion, from the inability of some of its more important stones to bear the necessary amount of pressure. And what terrible calamities in the falling of buildings, with cruel loss of life involved, do we have in consequence of stones which have been put in without sufficient trial."

"And partly, too, from the manner of building," suggested Ernest.

"Yes. A church was built in Constantinople, with the idea of rivalling in grandeur the Roman Pantheon, and twice its great dome fell in before the architects understood the principle of building a dome sufficiently well to succeed."

"Was the dome a part of the Doric or Grecian architecture?" asked Ernest.

"No; it is supposed to be wholly of Roman or Etruscan invention. The Greeks seem to have had no knowledge of it. It was introduced by the Turks into India, and became common in that country.

"But to return to building material. Some stones are put into buildings, which appear well to the sight, and which none but experienced workmen would reject, that, after all, are only 'composition stones,' and would come out very poorly if subjected to a severe test."

"What are composition stones?" asked Olive.

"They are *made* in different ways. Sometimes they are formed by pressing certain plastic substances in moulds. Potters' and different varieties of clay are baked, and subjected to other hardening processes, until they bear some resemblance to stone. Of course these products would never be thought of for foundation stones."

"Can any of you tell me, now, how Christ, our great Foundation Stone, was 'tried'? What was the pressure brought to bear upon Him?"

There was, for a little, no answer. Ernest felt the full force of the question, but said nothing.

"I suppose it was our sins," said Mary, at length.

"Yes; our sins. Think of it a moment. The sins of the world! 'Not for our sins only, but the sins of the whole world.' Can any of you quote a text proving this?"

"Behold the Lamb of God, who taketh away the sin of the world," said Olive.

"There is one in the third chapter of John," said Ernest. "For God sent not his Son into the world to condemn the world, but that the world through Him might be saved."

"Those are both very good ones. It is said in the fifty-third chapter of Isaiah, 'All we, like sheep, have gone astray, and the Lord hath laid on Him the iniquity of us all.'"

"That gives the idea of bearing, too," said Ernest.

"He bore our sins in His own body on the

tree. This is the way He was tried. Was ever weight like that? Was ever pressure so great?"

"I should like to ask," said Ernest, "how it was that He bore this weight. He was sinless, and how could He feel the pressure of guilt?"

"Let me try to explain this," said Mrs. Ewing, "for you could scarcely ask a more important question."

"It has puzzled me sometimes," said Olive.

"Did either of you ever feel very sorry for anything you had done that was wrong?"

"Yes," said Olive; "ashamed, and worried, and sorry."

"I am glad to hear you say that so frankly; but did you ever feel some special wrong you had done to be a very heavy *burden?*"

"I've had doing wrong make my heart ache," said Mary; "but I would do better than usual for a while, and then I would forget about it, and be as happy as ever."

"But after you have been 'as happy as ever' all day long, have n't you sometimes, in the quiet evening, had a sudden recollection come back to you of that wrong thing, that made your heart-ache come back too?"

Mary blushed, for she knew this was very true. Even she was not too young to know what a sharp thorn in the softest pillow a little sin could make.

"Yes, I have," she said.

"I have heard father say," said Olive, "that, when he was only four years old, he once wet his pillow with tears, crying almost all night, because he had helped, that day, to kill some little birds. He did not think, at the time, how wrong it was: but in the night he said it seemed as if he could not live and bear the weight of that sin unless God would forgive him. So he prayed very earnestly, until he felt the burden roll off his little heart, and then fell asleep."

"That sin was a part of the weight Christ bore," said Mrs. Ewing. "All the sin that is forgiven, all that sinners have a mind to cast upon Him, all they do not bear themselves of their own choice, He bears. Suppose, instead of, as Mary says, 'forgetting after a little while, and feeling as happy as ever,' your heart should keep on aching harder and harder. Perhaps you have been very angry; angry even with some very dear friend, perhaps with your mother, — that might be so, for children's hearts have some very dark spots of sin in them that need to be washed away, — and perhaps you said some word or words then which an hour afterwards began to trouble you very much. You could not bear to think you had said just that word. You tried to turn it, and twist it about, and make out to yourself that you really did not say just that very wicked thing. And worse than all, perhaps you told a falsehood if questioned kindly about it afterwards, and said you did not say it exactly as you did. But all this only added to the pressure of guilt on your heart. You could not help the matter any by trying to deceive others, or trying to deceive yourself. Something spoke louder than your own voice, and kept saying, over and over, and would be heard, 'Yes, you did say just that!' And so the burden grew heavier and heavier, until you felt as if a great stone was rolled upon your heart, and would 'crush it to powder' You could not bear it, you felt, any longer.

"Now just suppose that all your other sins, every wrong word, every harsh, unkind word, every slanderous word, yes, even every false word you had ever uttered, should rise before your mind with the clearness of this one sin, until the burden forced you at last to the cry of the publican, 'God be merciful to me a sinner!' then you would have some idea of the 'wrath of God' manifested to one soul; and then think of each soul since Adam burdened with this weight of sin, and no way to get rid of it themselves.

"Then Christ comes! He is omnipotent, and

can bear it; and so He takes the burden of your guilt, and the whole burden of every man's guilt, and bears Himself the wrath of God for it all; then you can have some idea of what it is to 'bear the sins of the whole world.' Christ offered to bear this. God accepted the offer. He saw the agony and bloody sweat long before the body of Jesus was bowed in Gethsemane. He heard the bitter cry, 'Why hast thou forsaken me?' and the later cry, 'It is finished!' long before the body of Jesus hung on the cross of Calvary. As soon as Christ said, 'Lo, I come: in the volume of the book it is written of me, to do Thy will, O God,' — as soon as He said this, God saw in Him a 'tried stone,' a sure foundation stone. And He has borne through the ages the test of all the pressure of guilt cast upon Him by weary mortals, who, falling under the load of their own heavy burden, have asked Him for relief. Millions have tried Him as their burden-bearer, have built upon Him as their only hope, and are to-day safe with Him, — safe from the curse which sin brings, because they trusted in Him who was 'made a curse for them.' 'Unto you who believe He is precious.' This is as much as we can go over to-day, as our hour is more than gone.

"For the next Sabbath I will ask you each to find some text which speaks of those who build upon this 'Stone,' as themselves being stones to be built up into a temple of Christ."

CHAPTER III.

PRECIOUS STONES.

"IN looking for verses about building stones," said Ernest, the next Sabbath, "I have been surprised to see how often this comparison about stones is used in the Bible."

"I have thought of the same thing," said Olive. "I began with Genesis, and looked along carefully to see what was the first mention made of stone."

Ernest laughed, but his mother kindly asked where and what it was.

"I suspect it was the stones in Aaron's breastplate," said Ernest.

"No, it was n't," said Olive; "it was long before that."

"Where, then?"

"In the very second chapter of Genesis. It says, speaking of the garden of Eden, 'And the gold of that land is good; there is bdellium and the onyx stone.'"

"Onyx was one of the kinds of stone in the breastplate, was n't it?"

"Yes; but another kind of stone is named before those in the breastplate," said Olive. It says, in the twenty-fourth chapter of Exodus, 'And they saw the God of Israel; and there was under his feet as it were a paved work of a sapphire stone.'"

"As it were," said Ernest; "that was n't stone."

"No; but it showed that they knew of the sapphire stone, at least."

Ernest smiled again; but Olive said,—

"I can't help it. I do n't see why we should n't be as much interested in Bible stones as in Bible trees and flowers; and a great deal is said about those."

"I think Olive is right, although we were going to talk more especially about building stones," said Mrs. Ewing. "I am glad, Olive, that you noticed these first mentions of stone."

"And perhaps they may come into the building stones," suggested Olive.

"They do n't build much with such costly stones," said Ernest. "You do n't suppose even palaces are built of precious stones, such as agate and onyx, do you?"

Olive was not going to give up entirely, and the way Ernest emphasized the word "*much*" made her secretly glad that she was not so wholly wrong as he supposed. She had, in truth, looked a little more carefully over her stone texts than Ernest; so she said,—

"I think when people make every preparation to do a thing, they generally do it. So I think precious stones were used about buildings in *some* way, if not to be put in the walls. I noticed, in the account of the things that David prepared or got together toward building the temple, that there was not only 'gold,' and 'silver,' and 'brass,' and 'iron,' and 'wood,' but" — and here Olive opened to where her little scarlet book-mark was placed, in First Chronicles, and read the last part of the second verse of the twenty-ninth chapter, with quite a little emphasis: —

"'Onyx stones and stones to be set; glistening stones and of divers colors, and all manner of precious stones, and marble stones in abundance.' And in the same chapter it says, a little further along, —

"'And they with whom precious stones were found gave them to the treasure of the house of the Lord.'"

"I wonder what part of the temple these precious stones were used for?" asked Ernest.

"In the fifty-fourth chapter of Isaiah, at the eleventh and twelfth verses, you will find these words," replied his mother: —

"'O thou afflicted, tossed with tempest and not comforted, behold I will lay thy stones with fair colors, and lay thy foundations with sapphires. And I will make thy windows of agates, and thy gates of carbuncles, and all thy borders of pleasant stones.'"

"They would have no need of stained glass, with such material," said Ernest. "I wonder if these precious stones were really used in buildings?"

"Of course there is much that is figurative in these verses, but no doubt they were used. They are not only mentioned in Olive's verse from Chronicles as a part of the material which David had collected for the building of the temple, but a little further on is a verse which Olive did not chance to see, I presume, where it is expressly said (2 Chron. 3:6), 'And he garnished the house with precious stones, for beauty.'"

"I'm so glad," said Olive, turning to the verse in her own Bible. "I wanted to be sure about their being in."

"What kind of stone is the onyx?" asked Mary.

"It is a variety of chalcedony, which is a translucent species of quartz. When it is found in stripes, it is called agate, and when the stripes run in horizontal lines across the stone, it is called onyx."

"Why is it called chalcedony?" asked Olive.

"Because found in Chalcedon, in Asia Minor. There are at least five varieties. One kind, found in the East Indies, Italy, and some other countries, is of a bluish-white color, lined with broad, white streaks; another, with the same ground, snowy-veined; a variety which is red-veined, called sardonyx, because supposed to be a mixture of the sardius and onyx; a kind similar to jasper; and a brown onyx, veined with blue. Pliny tells us there were onyx-marble quarries in Arabia."

"What was 'bdellium'?" asked Ernest. "That was mentioned first, before onyx."

"It is supposed to have been a precious

stone,—some have thought the same as beryl, but others have taken it to be a resinous gum; but, being named with gold and onyx, it was probably more precious than a gum, and we will think with those who call it a precious stone. Manna was said to resemble it in color, in Numbers 11 : 7. And as that was described as looking like 'hoar frost,' it was probably a clear or whitish stone."

"The onyx was the stone for Joseph's tribe, in the breastplate," said Ernest. "I noticed that, in reading over the description of it this morning."

"How could you tell?" asked Olive.

"Because there were four rows of stones, and three in a row. And it says the stones were named with the names of the children of Israel, according to their names. The breastplate was about ten inches square, and the stones were set in gold, in this way:—

A sardius, a topaz, and a carbuncle.

An emerald, a sapphire, and a diamond.

A ligure, an agate, and an amethyst.

A beryl, an *onyx*, and a jasper.

And if these corresponded to the twelve tribes, you can see that it would be,—

Reuben, Simeon, Levi;

Judah, Dan, Naphtali;

Gad, Asher, Issachar;

Zebulon, *Joseph*, Benjamin."

"It must have been a very beautiful ornament," said Mary.

"It was probably the first piece of mosaic ever made,—God's mosaic," said Mrs. Ewing.

"What did it represent?" asked Olive.

"It was worn by the high priest on his breast, fastened to the girdle of his ephod, and showed that he carried the twelve tribes, as it were, upon his heart before God. As the high priest represented them, so Christ bears His church upon His heart, presenting them before God by His intercessions. In that beautiful and glowing description of the New Jerusalem, given by John, in the twenty-first chapter of Revelation, it says, 'The wall of the city had twelve foundations, and in them the names of the twelve apostles of the Lamb. . . . Her light was like unto a stone most precious, even like a jasper stone, clear as crystal; and had a wall great and high, and had twelve gates, and at the gates twelve angels, and names written thereon, which are the names of the twelve tribes of the children of Israel. . . . And the building of the wall of it was of jasper, . . . and the foundations of the wall of the city were garnished with all manner of precious stones;' and their order is given in this way: 'jasper, sapphire, chalcedony, emerald, sardonyx, sardius, chrysolite, beryl, topaz, chrysoprasus, jacinth, amethyst.' This is God's mosaic again, a glimpse of wonderful beauty, caught by John, within the veil, which yet remains for all who build upon Christ, the corner stone. We shall realize, if we are ever permitted to enter in through the gates of pearl into the city, that 'eye hath not seen, nor ear heard,' what God hath prepared for those who love him."

"There was another very beautiful thing besides the breastplate," said Olive, after a moment, "which God ordered to be made for the priest to wear."

"The shoulder stones?" said Ernest.

"Yes. These were two onyx stones. They must have been pretty large, for six names were to be engraved upon each one. They were, it is said, to be for a memorial. What did that mean?"

"The same as in the breastplate; a constant reminder to him of the children of Israel, whom he was to remember earnestly and continually in prayer before God."

"How beautiful the idea of these 'memorial stones' was!" said Olive. "I never thought of them before except as ornaments, from their own beauty."

"Yes; the symbol was more beautiful than the ornament, as much more as pure love is better than pure gold or crystal. It gives us an early idea of God's great heart of love, as it has been revealed to men all along the ages. The memorials which Christ bears for us before His Father are the scars in His hands, and feet, and side."

"Is that what is meant in the verse of the hymn you like so much?" asked Olive:—

"'Five bleeding wounds He bears,
Received on Calvary;
They pour effectual prayers,
They loudly plead for me;
Forgive him, oh, forgive, they cry,
Nor let a ransomed sinner die.'"

"Yes; just it," replied her mother.

"I should like a 'memorial stone,'" said Olive, "if it would only help me to remember some things. I should like a red jasper, set in gold, for a ring."

"Is jasper ever red?" asked Ernest.

"Yes; it is found in various colors, which comes probably from a mixture with earth. It has been said, from this, that Christ might be said to be represented by jasper, as being 'at once the Lord from heaven and the fruit of the earth.'"

"I do not think there could be any harm in a 'memorial stone,' since God permitted them," said Olive.

"No, if we did not come to think more of them than of what they were selected to remind us of."

"I saw something the other day," said Olive, "which was beautiful as a 'memorial stone,' between two friends. It was Elsie Hartwell's ring. It had six stones in it, and they were set in such order that their initial letters spelled the word 'Regard.'"

"I never heard of such a thing," said Ernest.

"Well, I saw it, and there was first a ruby, and then an emerald, a garnet, an agate, then another ruby, and last, a diamond."

"R-e-g-a-r-d," said Mary; "what a very pretty idea!"

"Do you suppose," asked Ernest, "that

precious stones were really built in, in any part of the temple? I should like to know where those 'glistening stones' came in, 'for beauty.'"

"We know they were used in palaces in the East. Even the walls, which were themselves built of the purest and choicest marble, were sometimes inlaid with mosaic of precious stones. There were elegant borders, finished with festoons of fruit and flowers; the fruit composed entirely of precious stones, their own colors forming the color of the fruit and flowers. It may have been so in that gorgeous temple of St. Mark's, of which I read you, that 'range of glittering pinnacles, mixed with white arches, edged with scarlet flowers;' these flowers may have been carbuncles or rubies. At any rate, we know that it was sometimes done. Cornelian, agates, rubies, and other precious stones were so inwrought, in beautiful patterns, and in ornamental mouldings, as to appear as if carved and painted, rather than mosaic."

"Like the flowers in your Sorrento portfolio," said Mary; "inlaid, I mean, like that."

"How wonderful!" said Olive; "to think of an immense building, ornamented with precious stones, in walls of marble!"

"Do n't you remember, Olive," said Mary, "when we read the story of Esther, in our Golden Hour, once, how delighted you were with the account of the beautiful palace of the king — I can't think of his name —"

"Ahasueras."

"Yes. Where the beds of the palace were of gold and silver, and stood upon an elegant pavement of marble, of ever so many colors."

"Yes; 'red, and blue, and white, and black marble.' I remember it; for it made me think of that box Dr. Gerald showed us, that he brought from Rome. The cover was in squares of mosaic, and each piece was of a different colored, polished marble."

"Whole palaces were sometimes built in a sort of mosaic, in the East, though not of precious stones," said Mrs. Ewing. "I have read of buildings in alternate rows, or layers of black and white marble."

"I should like to see some of those magnificent palaces," said Ernest.

"You might be disappointed in them, certainly in some of which I have read. In Constantinople, hundreds of years ago, were temples and palaces of this description, where the material was exceedingly costly, precious stones abounding in endless mosaics; but the designs themselves were rude, and the work poorly executed. Here is a description of one of them, which I will read you. 'Toward the east, St. Sophia displayed its glittering domes, its hundred columns of jasper and porphyry, its precious marbles, veined with rose, striped with green, and starred with purple, whose saffron, snowy and metallic tints commingled as in Asiatic flowers, among balustrades and capitals of gilded bronze, before a silver sanctuary, facing a tabernacle of massive gold;

near golden vases, incrusted with gems, and beneath innumerable mosaics, decking its walls with lustrous stones and spangles of gold. The characteristics of this church, as of the entire city, were disorderly accumulation and unintelligent wealth. . . . People sought not beauty but bewilderment.' Of course this was not always so. With taste, the use of precious stones in buildings must have been very ornamental."

"It seems like a fairy tale, — the very idea of such palaces," said Ernest.

"If you were going to help build such a palace," said Mrs. Ewing; "suppose you were to be built in yourself, which would you rather be, an agate, to be used as a window to admit the light, a flashing carbuncle, to be set in gold, in clusters of fruit, or some dull, rude stone, fitted into an obscure corner?"

"I would be glad to be in at all," said Olive, answering, as she sometimes did, in Ernest's place, and occasionally somewhat to his relief.

"That is a good answer. We should be glad to 'be in, at all;' but if we can be in as 'polished stones, after the similitude of a palace' just as well, perhaps we should give greater pleasure to the great King. We are safe if we are in the true temple, and upon the true foundation, *at all*, no matter what niche we fill; but we may have regard to beauty as well as safety. The main thing is to be built on the right foundation. Some stones are very highly polished and very valuable and elegant, but if not built upon the true foundation stone, — if left in the quarry, by themselves, or if built up on some composition stone, that is neither 'tried,' nor 'sure,' nor 'elect,' nor 'precious,' however beautiful they may be for a while, they will only be part of a beautiful ruin in the end. And this only brings me now to the place where we were going to begin this hour. We shall only have time to hear your texts; and we will talk about them next Sabbath. Have you one, Ernest?"

"I found two verses in 1 Peter," said Ernest, "which suited so exactly that I felt surprised as I read them. 'To whom coming, as unto a living stone, disallowed indeed of men, but chosen of God, and precious: *ye, also*, as lively stones, are built up in a spiritual house, an holy priesthood, to offer up spiritual sacrifices, acceptable to God by Jesus Christ.'"

"That is a very beautiful text. '*Ye, also*.' Not only Christ, the foundation, but Christians, the stones, built upon it."

"Mine," said Olive, "comes in with the verse you said you wished we would always remember,—'other foundation can no man lay than that is laid, which is Jesus Christ.' It says, right after that, in 1 Cor. 3 : 12–16, 'Now if any man build upon this foundation, gold, silver, precious stones, wood, hay, stubble, every man's work shall be made manifest; for the day shall declare it, because it shall be revealed by fire; and the fire shall try every man's work, of what sort it is. If any man's work abide which he hath built thereupon, he shall receive a reward. If any man's work shall be burned, he shall suffer loss, but he himself shall be saved, yet so as by fire. Know ye not that ye are the temple of God?'"

"You see from that," said Mrs. Ewing, "that you may use very different material, and it may receive very different degrees of polish, — may even be stubble or wood instead of stone at all; but if built on the true foundation, even of the poorest material, you may be safe."

"What are the 'precious stones' in *this* temple?" asked Olive.

"First of all, Christ, the foundation, is a 'precious stone.' After this, I suppose, they represent the most perfect Christians. Those who are hewn by affliction, it may be, and polished by prayer, until their resemblance to their great Pattern is seen by others most clearly; these are the 'precious stones.' Those who are content with a certain degree of faith, and satisfied with enough hope to render them free from fear, but who have not love enough to make their lives beautiful, in the way of self-denial and noble deeds, may find some place in the temple, and may be enabled to 'stand' in the day when every man's work shall be made manifest; but it will be 'so as by fire.' It will not be an 'abundant entrance' into the 'New Jerusalem.' This forms an incentive to live such a life, that the beauty of it may be seen and recognized by all beholders, that you may be known as 'living stones,' 'precious stones,' an honor to the temple. Have you a verse, Mary?"

"Him that overcometh will I make a pillar in the temple of my God."

"That is a beautiful one. A real support; no hay or rubbish there; but a most valuable stone, even a *pillar* in the temple," said Mrs. Ewing.

"What does 'overcometh' mean?" asked Olive.

"I suppose it means more than merely to have faith sufficient for salvation; to 'overcome' the difficulties that will and must beset us in our spiritual life. I have seen persons who were 'pillars' for me, on account of their beautiful constancy of love in time of trial; their resistance to envy and to pride; their willingness to deny themselves for others, overcoming self, and overcoming sin, by living near to Christ, and depending wholly and constantly on Him for wisdom and strength. Such persons are, by their consistent life, an honor to the church; and more, a strength, a pillar, which God himself delights in."

"And what was your text, mother?" asked Ernest.

"It is found in the last four verses of the 2d chapter of Ephesians. 'Now, therefore, ye are no more strangers and foreigners, but fellow-citizens with the saints, and of the household of God; and are built upon the foundation of the apostles and prophets, Jesus Christ himself being the chief corner stone. In whom all the building, fitly framed together, groweth unto an holy temple in the Lord. In whom ye also are builded together, for an habitation of God through the Spirit.' These are very wonderful words. These 'living stones' were not always such; not until brought and placed upon the 'chief corner stone;' brought from among 'strangers' and 'foreigners,'—once no better than those who shall never be built in at all, because they will accept no place in this glorious temple. Their 'confusion' will be as great as was that of those who builded their tower of brick, which they hoped would reach unto heaven. 'Other foundation can no man lay.'"

"I see you do not mean us to forget that," said Ernest.

"No; it is the groundwork of all true hope. There is no sadder sight to me than to see men trying all their lives to get around this one great central fact; making themselves, with great care, into hewn stones, polished stones; but which can never become 'living stones,' because refused contact with the living foundation stone. Carried sometimes immense distances, and fitted with great care upon a false foundation, when the rains descend and the floods come upon them, the fall will be only the greater for all this seeming care and polish. 'The Lord knoweth them that are His.' No man can judge another. If one is upon this true foundation, no opinion of his fellow-man can shake him off into destruction. If he is not built upon it, no opinion of theirs can place him there. Honor, applause, eulogy, however great and flattering, will avail nothing. The foundation will remain true, whatever is built upon it, really or apparently."

"I was reading the other day," said Ernest, "of a very old cathedral in England which was being restored."

"Some historian had said years before that he thought it had originally been an old Roman temple; but this was doubted, until, in taking down the tower, they discovered under it 'real, genuine Roman masonry,' showing that it had been built on a Roman foundation; and afterward, in removing rubbish from an underground cell or crypt, they found several sculptured mitered heads, of fine Roman workmanship. So the foundation was thus proved to be Roman beyond a doubt, in spite of all that had been reared above it. We may find rubbish on this foundation; but we shall not have to destroy the whole temple to prove the work. There are 'sure stones' added daily to this building, and the time is hastening when it shall be completed, — when the top-stone shall be laid, and the shout of joy shall arise, 'Grace, grace unto it!' The prophets and apostles, from first to last, have been built upon this stone; and the millions who have tried it since the New Testament record closed, will never be put to shame."

"Was not Judas an instance of one upon whom this 'stone fell'?" asked Olive.

"Yes. His guilt was greater than he could bear. He had no foundation himself, because he sought to destroy the 'foundation' of a world. For our next hour we will talk about some of the 'living stones' which have been built into this temple during the centuries since Christ."

CHAPTER IV.

LIVING STONES.

"THE builders had refused the Stone. They had rejected Christ; condemned and crucified Him. His body had been laid in the sepulcher amid sweet spices, prepared by weeping friends, who yet had not faith to interpret His prophecy: 'I will destroy this temple, and in three days I will raise it again.' There was a stone at the door of the tomb, and a seal upon it; and a guard was set. But He had not only power to 'lay down His life,'—He also had power 'to take it again.' No man saw the breaking of that seal; but when Mary Magdalene came early, 'while it was yet dark,' on the first day of the week, she saw that the stone had been rolled away, and ran, and said to the disciples, 'They have taken away my Lord, and I know not where they have laid Him.' And then He revealed Himself to her. The disciples had not understood Him in all that He had told them of His death and resurrection, but now they saw 'the Lord's doings,' and it was indeed 'marvelous in their eyes.' He had now become 'the Head of the corner.' You may turn to the 2d chapter of John, Olive, and read what Christ said to the Jews about his death, commencing with the 19th verse."

Olive read: "Jesus answered and said unto them, 'Destroy this temple, and in three days I will raise it up.' Then said the Jews, 'Forty and six years was this temple in building, and wilt thou rear it up in three days?' But He spake of the temple of His body. When, therefore, He was risen from the dead, His disciples remembered that He had said this unto them, and they believed the Scripture, and the word which Jesus had said."

"So you see it made no difference with the cause of Christ, that 'the kings of the earth stood up, and the rulers were gathered together against Him.' They had only carried out the counsel of God. There is nothing that can stand against this. And God will see that all who oppose Christ's kingdom shall be made to know, either in love or in wrath, that 'this same Jesus whom they crucified is both Lord and Christ.' Soon after the ascension of Christ, a wonderful proof was given of this, in the conversion of three thousand in one day. The building stones—living stones—for the new temple began to come in from every quarter. The rich and the poor, the ignorant and the learned, all furnished them; and while there remained 'despisers' to 'look on, and wonder, and perish,' the work still went on. David says, in the 127th Psalm, Except the Lord build the house, they labor in vain that build it;' and it is no less true, that if He be the Master Builder, they will labor in vain who may try to destroy the work He is carrying forward. David felt this, when he prayed, in the 51st Psalm, 'Build thou the walls of Jerusalem.' David seemed to rejoice in this figure of the church as a building. I have been interested in noticing how often he uses these expressions: 'When the Lord shall build up Zion, He shall appear in His glory.' 'The Lord doth build up Jerusalem; He gathereth together the outcasts of Israel.' And so with the prophets; running all through their writings is this figure, and the same truths are taught by it. There is one so beautiful in Zech. 6 : 12, 13, that you may read it, Ernest."

"Behold the man whose name is The Branch; and he shall grow up out of his place, and he shall build the temple of the Lord; even he shall build the temple of the Lord; and he shall bear the glory, and shall sit and rule upon his throne."

"And in the 15th verse of the same chapter,

you see that the Gentiles are to help in this building."

Ernest read: "And they that are far off shall come and build in the temple of the Lord, and ye shall know that the Lord of hosts hath sent me unto you."

"Those are very striking words," said Mrs. Ewing, "and I think that Paul had them in his mind when, after Christ had come, and the visible foundation had been laid, and he had gone to work in the Gentile quarry, and was trying to work out 'living stones' for this building, he said, in speaking to these Gentiles, in Eph. 2: 'But now in Christ Jesus, ye who sometimes were far off are made nigh by the blood of Christ; for he is our peace who hath made both one, and hath broken down the middle wall of partition between us. . . . Now, therefore, ye are no more strangers and foreigners, but fellow-citizens with the saints, and of the household of God; and are built upon the foundation of the apostles and prophets, Jesus Christ

Himself being the chief corner stone. In whom all the building, fitly framed together, groweth unto an holy temple in the Lord. In whom ye also are builded together for an habitation of God, through the Spirit.'"

"Paul uses the very same expression as Zechariah," said Olive; "ye who were 'far off.'"

"Yes; and what a wonderful truth is clearly taught us in these words. All may come and be built into this temple; not only the Jew, but the 'outcast.' One may be made to serve just as useful a purpose as another, if only all are on the same foundation. Who would not make sure of a place in this building, which shall stand all the shocks of time, and survive all commotions and changes; which shall stand firm when every other temple shall become as did the great and glorious temple of the Jews, so utterly ruined that not '*one stone* was found left upon another'? And who is able to do this for us? Even little Mary can tell. You may read a promise, Mary, in Acts 20 : 32."

"I commend you to God, and to the word of His grace, which is able to build you up, and to give you an inheritance among all them which are sanctified."

"Yes, God is able to do this; and if we ask Him, He will do it for us. He will make all of us 'living stones' in this great temple,—precious stones, polished stones,—and no one can take us out of His hand.

"Do you remember, Olive, that elegant stone mansion—almost a palace—that your father pointed out to you last summer, when we were traveling in New England, and which was left unfinished, with its towers uncapped, and without roof or cornice, exposed to the weather, and already growing dingy and moldy as it stood?"

"I do," said Ernest; "somebody's 'Folly,' they called it; and I remember some one quoting the text, as we passed it, 'This man began to build, and was not able to finish.'"

"Yes; I remember it, too," said Olive. "It seemed so sad to me. I did not feel like laughing, at all."

"Nor I," said Mrs. Ewing. "It was very suggestive to me at the time, and has been more so as I have thought of it since. But will this be the case with any building which the Lord shall undertake?"

"No," said Mary.

"Why not?"

"Because there would n't be any reason why *He* should ever stop," said Mary.

"That is the very best of all reasons. He has everything at his control. He can speak a world out of nothing. Even this foundation stone, it is said, was 'cut out of a mountain without hands.' God prepared it in His own way. And He can of course prepare and bring from whatever quarter He chooses every lesser stone to be built upon it; and He will see that not one shall be broken, or marred, or left behind, in the place where it was prepared, but that all shall be safely built in, so as to stand firm and secure as the Everlasting Rock beneath it. And this we have seen all along the ages.

God has not left this building to go on without witnesses to its constant growth, and to the stability and perfection of it, so far as it has advanced. I said we would talk, this afternoon, of some of those who have already been built upon this foundation. Can you tell me, Ernest, who were some of these 'living stones' in the first century?"

"Those mentioned in the New Testament would come first, would they not? You spoke, a while ago, of the three thousand who were brought in on the day of Pentecost," said Ernest.

"This was but the beginning," replied Mrs. Ewing.

"Stones for the temple were daily brought in. In the same chapter which tells of the three thousand, we are told, 'And the Lord added to the church daily of such as should be saved.' And when the Jews, who had always thought themselves the only fit building stones for God's house, murmured, and questioned about this new material, and imprisoned Peter and John for their share in the work, asking 'by what power or by what name they had done this,' Peter told them boldly that it was Christ's own work; and added, '*This* is the stone which was set at naught of you builders, which is become the head of the corner.' And although they marveled, yet they thought to put a stop to the work, and commanded them 'not to speak or teach at all in the name of Jesus.' But their reply was, 'We can not but speak the things which we have seen and heard.' So these men, bold in Christ, were released; and when they went back and reported this to their friends, the reply was, 'Lord, thou art God!' Their own faith was strengthened, 'and believers' (true, 'living stones') 'were the more added to the Lord, — multitudes, both of men and women.'

"Again, these apostles were imprisoned for preaching Christ, but the 'angel of the Lord' opened the prison doors and brought them forth, and said, 'Go, stand and speak in the temple to the people all the words of this life.' Do you think *men* could stop such a building as this? No wonder, when angels and earthquakes blocked up the way of the old Jewish builders, they 'began to doubt whereunto these things would grow.' One wise man among them gave them at length this excellent advice: 'If this work be of men, it will come to naught; but if it be of God, ye can not overthrow it, lest haply ye be found to fight against God."

"There was a great deal in that '*if*,'" said Ernest.

"Yes; and it has been well proved whether the work was of men or of God, by Gamaliel's own test. Eighteen hundred years have only added to the force of his words; for not a stone laid on this foundation has yet toppled from its place."

"Did none of these old Jewish builders change their minds, and fall in with the new work?" asked Ernest.

"Yes; 'the word of God increased, and the number of disciples multiplied in Jerusalem greatly; and a great company of the priests were obedient to the faith.' God was not satisfied wholly with small stones. He could select a pillar here, and another there, just as he needed them. He can make 'the weak confound the mighty,' and he can also use the mighty, as well as the weak. The 'priests' were the best builders among the Jews. They had received the 'memorial stones.' Their business was to take the charge of the temple. They were chosen men, sound in body, not lame, or blind, or deformed; sound in mind, of good judgment, consecrated by sacred rites, and anointed for their work. 'A great company of these' became pillars in the building reared on this 'tried' foundation. No matter how hard it was for them to change their opinion with regard to this rejected Stone; no matter if at first they came, like Nicodemus, 'by night,' to test him, whether he were to be refused or accepted; it was an easy thing for God to turn their hearts from doubt to certainty, from hatred to love, so that, like this same Nicodemus, they might be found, at last, bringing their costly tributes of myrrh and aloes to a living, instead of a dead, Lord and Christ. And in the hearts of those who accepted this stone God placed a firmness that nothing could move."

"One stone, in this century, I think we should notice particularly," said Olive; "for it was *laid in blood.*"

"What do you mean?" asked Mary.

"She means the first martyr," said Ernest,—"Stephen. I had been thinking about him."

"He was a very zealous worker in this quarry of the Gentiles. He was 'full of faith and power,' and did 'great wonders and miracles among the people.' So they tried to crush him, although 'they were not able to resist the wisdom and spirit by which he spake.' When he told them boldly of the Stone they had refused, they stoned him, and thought to destroy one of the fairest 'living stones' of the glorious temple which he was helping to build; but, look at this temple from whatever point we may, where shall we find a fairer block to-day than that inscribed with the name of the martyr, Stephen? In that moment, when his perfected faith caused his face to look 'as it had been the face of an angel,' God saw in him a polished stone, whose beauty the passing of ages should never mar."

CHAPTER V.

PILLARS.

"SAUL was another noble pillar of this first century," said Ernest, the next Sabbath afternoon.

"Yes; Christ became a 'precious' and 'sure' foundation to him, in a wonderful manner. He had before been to Saul so great a 'rock of offense,' that he had well nigh stumbled into fearful ruin; but God had 'selected' *him* also, saying, 'He is a chosen vessel unto me to bear my name before the Gentiles, and kings, and the children of Israel.' So completely turned about was Paul, so earnestly did he begin his work for Christ, that all who saw him 'were amazed, and said, Is not this he that destroyed them which called on this name in Jerusalem?' From destroying and 'tearing down,' he began to 'build up;' and he could only exclaim, as did Peter, when called to account for the change, 'What was I, that I could withstand God!' And so faithfully and zealously did he work for his new Master, that others caught his spirit, and saw, as he saw, that this was indeed God's work, and began to quote (as did James in Acts 15:15) from the Old Testament, to prove that this building was of God. Olive, will you read this quotation, and the place also in Amos where it was first written?"

"'And to this agree the words of the prophet; as it is written, 'After this I will return, and will build again the tabernacle of David, which is fallen down; and I will build again the ruins thereof, and I will set it up.' The verse is almost the same in Amos 9:11. 'In that day will I raise up the tabernacle of David, that is fallen, and close up the breaches thereof; and I will raise up its ruins, and I will build it, as in the days of old. That they may possess the remnant of Edom, and of all the heathen, which are called by my name, saith the Lord, that doeth this.' "

"I wonder how James came to think of this," said Olive, after carefully reading the quotation.

"It was because he was taught of God. These words of the prophet were spoken more than seven hundred years before Christ came. Amos looked forward, and saw upon what he was building; and James beautifully acknowledges this, and finds in it a confirmation of what his own eyes see and his own ears hear; that which even prophets and kings, though they 'believed,' yet 'desired to see and hear' more plainly. If we remember that God is 'the same yesterday, to-day, and for ever,' that with him it is one 'eternal *Now*,' it will not seem so strange, this quick linking of James, and all the New Testament saints, with the Old Testament prophets. With his eye they were often permitted to take these forward and backward glances, making the past and present, for the moment, as one. James closes his remarks, after this quotation, with these remarkable words: 'Known unto God are all his works, from the beginning of the world; wherefore my sentence is, that we trouble not them which from among the Gentiles are turned to God.'"

"It seems so long," said Mary, who had been trying to keep up her interest all the while, this building going on away back before Amos, I do n't know how far, and on, I do n't know how far. When I do anything, I like to see it done right off."

Ernest laughed. "That is the way with people who do *small* things," he said. "Somebody's teacher said the other day, if I heard aright, that a certain rose and butterfly would have been a good deal better painted, if it had not been finished pretty nearly as soon as it was traced. If you put on a second coat, even in water colors, before the first is dry, it makes rough work. *I* like to see things that can wait a while."

"Mary is a little girl, Ernest. There is a grandeur, it is true, in the idea of time in the completion of a great work. God never needs to be in haste. If man would see the result of his labor, he must necessarily work fast; this is, no doubt, why we look with a sort of awe on works that men begin, knowing certainly that they themselves will be in the dust long before they are finished."

"Like the palaces we read of, — some of the old French cathedrals, that were hundreds of years in being builded," said Ernest. "I read in a magazine, only a day or two ago, of work still going on upon the cathedral in Cologne, which was begun a great many hundred years ago. One tower of this old cathedral, it was stated, grew thirty feet in height, and another twenty, during the year of 1869. And not only the outside, but the inside work is also being still carried on."

"I was interested in seeing the same statement," replied his mother. "That cathedral was begun more than a thousand years ago, — eight hundred and fourteen years after Christ. It was burned once to the ground, and a new foundation laid in 1162, and it has been slowly creeping up through the centuries ever since. There was a legend that Satan was jealous of the work because it was so vast, and made a vow that it should never be completed. If so, it will probably do as much toward hindering the work as his vows and efforts with regard to God's great building,—as much greater than this cathedral as that is greater than the palace of a child's building blocks, — will hinder its completion. The top-stone is to be put on to the cathedral, doubtless. And more surely will the time come when the top-stone will be laid to this eternal temple, and when all the sons of God, beholding it, shall shout for joy. Satan may seem to triumph for a while; but he is chained, and God can as easily say to him as to the fretting sea, that tosses continually against the water-worn rocks, 'Thus far shalt thou go, and no farther.' The fairest earthly temples, however long in being builded, however strong in architecture, and perfect and elaborate in detail, from the solid foundation-marble to the small rose-shaped windows, wrought with the lightness and intricacy which give 'almost the airiness of lace,' shall fall before Time and all its changes; while this temple of God shall remain. Paul felt this when he said, 'I know in whom I have believed, and am persuaded that he is able to keep that which I have committed to him against that day.' This was the secret of his life of untiring work for Christ. *He* was safe for eternity so soon as he laid himself upon this sure foundation; and now, all he could do toward that building would be work for eternity. Every stone he could bring, large or small, unhewn or polished, common or precious, for this slowly ascending temple, was so much work done, never to be undone, —'laying up treasure' for heaven.

"It made no difference to him that those to whom Christ was still a stumbling-block, could not see that he was working wisely; no difference to him, that some said he was 'turning the world upside down,' so long as, by faith, his eye was fixed upon this 'temple not made with hands,' which was to be 'eternal in the heavens;' no difference to him that he was reviled at Ephesus when he spoke boldly in their vast temple, so long as many believed, and 'the name of the Lord Jesus was magnified.' One of the charges against him was, 'that the temple of the great goddess Diana would be despised' through his teaching that 'they be no gods which are made with hands.'"

"Was not the temple of Diana one of the seven wonders of the world?" asked Ernest.

"Yes; it was a very ancient and very magnificent temple. It was at least two hundred and twenty years in being builded. Many attempts had been made, from different reasons, to destroy it. Seven times it had been set on fire, but each time restored, and made larger and more beautiful. Its length was four hundred and twenty-five feet, and its breadth two hundred and twenty feet. Its roof was supported by one hundred and twenty-seven marble pillars, sixty feet high. Twenty-seven of these were richly and curiously carved; the rest were polished. It is said that each pillar was cut from one block of marble, and each was the gift of a king. And all this great work was for a goddess made of ivory, and ornamented with gold, whom the people believed to have fallen from the skies! About three hundred and sixty years before Christ, this temple of Ephesus was burned by a man named Erostratus, who, it has been said, having no good works to recommend him to notice, determined to seek notoriety by some great work of evil. This was on the same night in which Alexander the Great was born; and years after, when he became a successful conqueror, he offered to rebuild this temple, if *his name* might be inscribed upon its front.

The people rejected this with scorn, and rebuilt the temple with aid from all the Ionian cities. It was about fifty years after Christ that Paul visited Ephesus, and began to bring stones for an 'unseen temple' from among that people. The marble pillars of the temple of Diana have long since disappeared. A little Turkish village now stands upon the site of this once splendid city; and it is said that so thorough has been the work of destruction, that the site of its more prominent buildings, even of the temple of Diana itself, is now a matter of conjecture. A recent visitor to the place says, 'Here and there only does a fragment of a temple wall appear above the soil, while, usually, prostrate columns and scattered capitals, bases, and cornices, and squared stones, are the only evidences of the grandeur that was. Not a single erect column can be found in the whole city.' And another writer, in speaking of this utter destruction, says that 'the entire disappearance of so huge a mass as the temple of Diana can only be accounted for by supposing that the materials were carried away and incorporated into other buildings.' But however it was done, the temple is gone; and those who built it, and those who believed in the great goddess for whom it was built, and who shouted, when deriding Paul's work, 'Great is Diana of the Ephesians!' have gone with it. Its foundations were false, and when 'tried,' they were found wanting. But those 'living stones' which Paul brought from Ephesus remain, and will be seen and known when the work is finished; for it was to these very Ephesian converts that Paul said in his letter to them, from which I before quoted, 'In whom *ye* also are builded together for a habitation of God, through the Spirit.'

"Paul had already a glimpse within the vail. Already he saw the foundations of the jasper-walled city, — the jasper, and the sapphire, and the chalcedony, even to the amethyst. He saw them as John saw them (if with less open vision), and for this he too, like Stephen, was willing to add another blood-sealed stone to the Master's building. Nobly he 'overcame,' and that promise was made sure to him which Mary read,—'Him that overcometh will I make a pillar in the temple of my God.'"

CHAPTER VI.

ASBESTOS STONES.

"WHO were some of the living stones of the first century, besides those in the New Testament?" asked Olive, on the next Sabbath.

"I will tell you of some of them. We shall be interested, I am sure, to follow this building through the centuries up to the present time, although we may only be able to glance at a pillar here, a tower there, or a precious stone, it may be, suddenly sparkling from some mosaic ornament, along the uprising wall."

"I shall look out for the agates, and the carbuncles, and the amethysts, and the rubies," said Ernest.

"And I for the great marble pillars cut from one block," said Olive.

"And I for the little stones that 'fit in anywhere,'" said Mary.

"In looking out for the little stones that 'fit in anywhere,'" said Mrs. Ewing, "some have found bright jewels, 'prisoned as topaz-beams within a rough-garbed stone.' Where were the building stones to be brought from, after Christ, the foundation, was laid, Ernest?"

"From anywhere in the world, I suppose."

"What did Christ say that showed this?" asked Mrs. Ewing.

"I am not sure."

"He said, 'Go ye into all the world, and preach the gospel to every creature.' The world is the great quarry now; and the workers, though few, are busy and faithful; and however widely scattered, all keep their eye on the same great Master-builder, to whom each will render his account. There is, to-day, the sound of the ax and the hammer, blocking out stones for this temple in the long-sealed quarries of Spain, and the newly-opened ones of far-distant China and Japan. You may help to put in a 'polished stone' in this temple, sitting quietly at home, by giving to aid the workmen now on the ground. We must in some way attend to the command of the Master-builder to 'go into all the world;' and it will not do to say that to attend to the nearest need — that immediately about us — is obeying this command, unless we can make ourselves believe that our own town, or city, or country, is 'all the world.' But by giving even a small portion of what God freely gives us, we can literally go into all the world, and help furnish material for this great temple. It will be built, and completed, whether we ourselves are found upon it or not; or whether, being built upon it ourselves, we lay a single other 'block' upon the building or not. Let us not be blinded to the privilege of bringing many stones, and of acting ourselves as hewers and 'stone-squarers' for this great temple. And let us be careful to work willingly, knowing that the Lord loveth a cheerful worker, as well as 'a cheerful giver.'

"It has been said that 'no artist can produce the highest specimen of art, no architect the best specimen of building, except he have a love for his work. He must be *more* than *willing;* he must *enjoy* the labor.' So our work on God's temple, to be beautiful in His sight, must be the 'labor of love.' He will see more beauty in the gift of a cup of cold water from love to Christ, than in the offering of golden guineas, if done reluctantly or grudgingly, or from any motive but that of love.

"Peter, and Paul, and John, and the other disciples did a great deal of this work for others, so that, although we do not know how far they went in this direction, except what we learn from the New Testament, we do know that, during the first century, a great deal was accomplished for this building. And it was indeed marvelous, when we consider the workmen, — weak men, men in humble positions in life, many of them ignorant and unlearned men, — and also when we consider the work which they had to destroy, as well as that which they had to build. There was much to contend against in the prejudices and passions of men who had long been building on false foundations; but the disciples proved, through the power of God, 'mighty in pulling down these strongholds,' and taking from them material for their new work. Many sealed their testimony with their blood. Like the old builders of the temple wall, in the time of Nehemiah, 'they which builded on the wall, and they that bare burdens, with those that laded, every one with one of his hands wrought in the work, and with the other hand held a weapon. For the builders every one had his sword girded by his side, and so builded.' This did not save Stephen from stoning, or

James from the sword, or Paul from the headsman's ax, or Peter from the cross, or John from exile, or Ignatius from the fury of wild beasts at Rome. The Jews had fallen under the power of the Romans, and Nero persecuted the Christians for years in the most terrible manner. But although he could destroy earthly temples, and burn cities, and hinder the workers on the wall, he could not stop the work. Among the noted builders of the second century was Justin Martyr."

"Was he a martyr?" asked Olive.

"Yes, and that is why he was called by that name. He was not a rough, unhewn stone. He had been polished before he thought of being made a stone for the temple of Christ at all. He was a learned man and a philosopher. He not only spoke, but wrote books, for the truth, five of which still remain. From him we get some glimpses of the manner in which their worship was conducted in that early day, when, for fear of persecution, men met in caves, and in the tombs for the dead, as well as in private houses, for their worship. This may seem like building in the dark; but it was building, and seventeen hundred years have not destroyed the work."

"What did he say about their worship?" asked Olive.

"He said, 'On the day which is called Sunday, all, whether dwelling in the towns or in the villages, hold meetings; and the memoirs of the apostles, and the writings of the prophets, are read, as much as the time will permit; then, the reader closing, the president in a speech exhorts and excites to an imitation of those excellent examples; then we all rise, and pour forth united prayers; and when we close our prayer, bread is brought forward, and wine and water, and the president utters prayers and thanksgivings, according to his ability, and the people respond by saying, Amen; and a distribution and participation of the things blessed take place to each one

"Men met in caves, and in the tombs for the dead." Page 106.

present, and to those absent it is sent by the deacons. And those who are prosperous and willing give what they choose, each according to his own pleasure; and what is collected is deposited with the president, and he carefully relieves the orphans and widows, and those who from sickness or other causes are needy, and also those in prison, and the strangers that are residing with us, and, in short, all that have need of help. We all commonly hold our assemblies on Sunday, because it is the first day, on which God converted the darkness and matter, and framed the world; and Jesus Christ our Saviour on the same day arose from the dead.'"

"That is a beautiful account," said Olive; "so simple, and yet so good! I would like to attend just such a meeting."

"I would like some singing," said Ernest.

"Although he does not mention that as a part of the service, other early writers do; and perhaps he included it in the word 'Thanksgiving.'"

"How did Justin become a Christian?" asked Ernest.

"He had studied different forms of philosophy, but his mind was restless and dissatisfied; he wanted something better — a true foundation — to rest upon. He was a native of Sychar, in Samaria. He was one day walking on the sea-shore, when he chanced to meet an aged Christian man, in whose conversation he became at once interested, and by whom he was persuaded to read the Jewish Scriptures. This interview led, in the end, to his conversion. He afterward went to Alexandria, and to Rome; and wherever he went, like Paul, he was earnest in his work for Christ. He was put to death during the reign of the Emperor Antoninus, about A. D. 166."

"How was he martyred?" asked Mary.

"The manner of his death is not certainly known. Some think he was beheaded."

"It is sad," said Olive, "to think how many

lives have been lost in connection with this building."

"Yes," replied Mrs. Ewing; "but we must remember whose lives were lost. Not the martyrs', but theirs who caused them to perish. They are still living stones, while their persecutors are the ones who are lost. It is the despisers, who 'wonder and perish.'"

"Was not Polycarp a martyr of this century?" asked Ernest.

"Yes; he is said to have been a disciple of the apostle John. He suffered martyrdom only three years after Justin Martyr. He furnished many a stone for the true temple. He was educated by a Christian lady of Smyrna, and afterward became bishop of that city. During the persecution of Christians under the Emperor Aurelius, he was taken from the villa to which he had fled for security. He was carried to Smyrna, being treated with contempt and insult all the way, which he bore with great meekness. When urged to save his life by denying Christ, he replied, 'Eighty and six years have I served Him, and He has done me nothing but good; and how could I curse Him, my Lord and Saviour? If you wish to know what I am, I tell you frankly, I am a Christian.' The proconsul then said, 'Polycarp has confessed himself a Christian!' which announcement was the signal for his death. He refused to be fastened to the stake, saying, 'He who has strengthened me to encounter the flames will also enable me to stand firm at the stake.' He then offered a short prayer of thanksgiving, in which the sentence occurs, 'I praise Thee that Thou hast judged me worthy, this day and this hour, to take part among the number of Thy witnesses in the cup of Thy Christ.' Wonderful words! Only the great power of God could have worked this faith in a mortal man."

"His work was really 'tried by fire,'" said Olive. "He was a tried stone, a pillar of the church."

"A pillar of asbestos stone, almost," said Ernest.

"What is that?" asked Olive.

"It is a mineral upon which fire has no effect. There is a fine variety of asbestos, which is pure and lustrous, and has been compared to white satin. From this quality it has been called 'Amianthus,' meaning 'undefiled.' From the purity of the character of Polycarp, he might remind us of this wonderful stone."

"I should like to be such a stone," said Olive. "I like the name, and its meaning, the 'undefiled.'"

"You would n't like to earn it by being a *martyr*, like Polycarp and Justin Martyr," suggested Ernest, "even to be called the 'undefiled.'"

"That would not be necessary, or even possible," replied Mrs. Ewing. "It was not the *suffering* of Stephen or of Polycarp that saved them, only as that was a test of their true, genuine faith in Christ. They may have become more polished, better fitted for an important place in Christ's temple, by means of the trial; but many will be found among the 'undefiled,' in the day of final trial, whose path was not hard and thorny, and sealed with blood, but who walked, most of the way, in the sweet sunlight of God's love, by faith, His '*goodness*' having led them to repentance. You spoke of 'earning' the place of the 'undefiled,' Ernest. Can we earn a new and pure heart in any way?"

Ernest felt a flush upon his face, as he replied, "I thought it was something to be a martyr. I mean, I thought one who suffered so much would be looked upon with peculiar favor by God."

"Many mistake here. Those who think there is saving virtue in suffering forget that it depends wholly on the reason of the suffering. There is all the difference in this that there is between a martyr and a suicide. Judas suffered more than Stephen, but it did not prevent Christ saying of him, 'It had been good for that man had he never been born.' Neither was it the *suffering* of the thief on the cross that insured his salvation. He had already borne the cruel pain of the nails that fastened him to the wood, he was already the 'dying' thief, when by faith he lifted his filmy eyes to the Saviour with his soul-wrung plea, 'Lord, remember me when thou comest into thy kingdom.' It was not his suffering, but his faith, that brought forth the gracious and ever-wonderful reply of the suffering Saviour on the cross beside him, 'To-day shalt thou be with me in Paradise.' Christ saw in him a sinner who had — although at the last moment, and in the extremity of suffering — 'come to him;' and He was true to His own promise, 'And I, if I be lifted up, will draw all men unto me.' 'Whosoever cometh unto me I will in no wise cast out.' Christ was now literally 'lifted up,' and the poor, dying, crucified thief was the first to realize that 'as Moses lifted up the serpent in the wilderness,' that all who looked upon it might live, so 'the Son of man was now lifted up,' that whosoever looked to Him might have eternal life. Not whosoever suffered as He suffered, but whosoever looked to Him, should live. The other thief who was crucified at the same time suffered as much as the one who looked to Jesus, but the suffering did not save him. Remorse often leads men, as it led Judas, to inflict upon themselves greater suffering than the wrath of wicked men, with all their hellish inventions of torture, could inflict upon those for whom they secure a martyr's crown. We do not say that God does not look with a peculiar tenderness upon the martyr. But it is because of the love which it reveals, when he is tried even so as by fire, — even unto death, — and not found wanting."

"I never thought about suffering in just that way before," said Olive. "I heard old Mrs. Glenn say one day that 'she thought the Lord would be kind to her at last, because she had suffered so much all her life, in poverty and sickness, and so many ways.'"

"Not because of the suffering. Suffering

may sometimes be necessary to touch the heart, and lead it thus to look to the great Helper for the only true relief; but God is as well pleased with, and as ready to bless, those who allow His goodness to lead them to repentance. I know a person whose life has been very free from sorrow, and from bereavement and losses, and whom once I heard say that this was her daily prayer: 'Let Thy goodness lead me to repentance.' And when those who are Christ's children are permitted to suffer for Him, even though it be the suffering of a martyr, God gives them strength to bear it. If He does not appear in the midst of the furnace, and, in 'the form of the fourth,' keep the scorching flame from harming His chosen ones, He is still a joy and a strength in their souls all the way through. They hear Him whisper,—

> 'The flame shall not hurt thee. I only design
> Thy dross to consume, and thy gold to refine.'

The gold was there. It is not made, only refined, by the suffering."

"I'm sorry to think of the way Polycarp died, — sorry yet," said little Mary.

"Yes, my dear; the history of the martyrs is very sad. When I was a child, I could hardly bear to look into the old 'Book of Martyrs' in my father's library, and could only read it at all by saying to myself, 'That was over long ago.' But even their suffering was short compared with the 'glory which should follow.' Christ was 'tried,' and many of His followers have been tried with Him. They shall also 'reign with Him;' and then they may class even martyrdom among the '*light* afflictions' which have wrought out for them a far more exceeding and eternal weight of glory."

"Does n't that look a little like earning it? I mean the expression, 'wrought out for them,'" said Ernest.

"You might at first think so. And the faith, no doubt, was strengthened, and the love increased, by the trial, but not *made* by it. Hewing and polishing may add to the beauty of a 'living' stone; cutting and polishing may bring out the hidden beauty of the jasper, or onyx, or crystal, — may *reveal* what otherwise might have been passed by as dull and worthless; and this, we know, is often the case. But unless there is beauty in the stone, no amount of cutting or crushing will bring it out. Unless there is life in the 'living stone,' no amount of mere suffering will reveal or place it there, or so purify it as to place it among the 'undefiled.'"

"I've often broken open a perfectly dull-looking stone," said Olive, "and been more delighted to find sparkling crystals within, than when I have found larger and more brilliant ones in the geodes, which hinted beforehand, through their little round glinting openings, something of what I might expect to find inside."

"And I have experienced the same pleasure with regard to 'living stones,'" replied her mother. "Once, a slight touch, the merest attempt to see what was within, and when I really expected nothing, startled me by its sudden revelation. It was in the case of a person not only dull, but so wanting in understanding that her best friends scarcely thought her capable of judging between right and wrong. She could only read a few easy words, and make out the very simplest text of Scripture, with the help of a friend and great pains-taking, when, one day, happening to be left alone in the room where she was sitting, humming to herself as was her constant habit, I asked, rather abruptly, 'Mary, have you ever thought about Jesus Christ as a Saviour from sin, — as one who would be your friend?' Never did a dull stone fly open at the stroke of a hammer to show more unexpected beauty in a moment, than this question revealed. With a lighting up of the whole countenance, and a beam of intellect in the eye which surprised me, the answer was, 'I'm glad you asked me the question. Nobody has ever asked me before; but for a good while I have loved the Saviour. I know I love Him.' 'How are you sure?' was all I could ask, in my sudden surprise. 'I think of Him all the day. I've read a few of His kind words; I've asked Him for a new heart; and I know I love Him. I should not be at all afraid to die to-night. I should go right to Him. I know He would be there, ready to receive me. I should not be at all afraid.' Tears were rolling silently down her cheeks, and if ever love, and the triumph of clear, simple faith, shone out upon a face, I saw it in hers.

"Not long after, she was tried. Sickness came, and death soon followed; but all who saw her then knew that 'Jesus was there, ready to receive her.' She was 'washed white in the blood of the Lamb.'"

"There is a text in Job," said Olive, "which I have often thought of with pleasure; and that makes me think of it,—'His eye seeth every precious thing.'"

"Yes; it is a beautiful thought. God never stumbles accidentally upon the hidden treasures, whether of natural beauty or of the heart. He sees them all. What we may pass by with scorn, because we do not know how or where to apply the magic touch that shall reveal its hidden beauty, is ever open to His eye. The whole of that verse, — that and the one before it, — is very beautiful: 'He putteth forth his hand upon the rock; he overturneth the mountains by the roots. He cutteth out rivers among the rocks; and his eye seeth every precious thing.' That is the way in which He works among the 'living stones' in the world's great quarry. He touches the flinty heart, and streams of love gush forth; even rivers flow from among the rocks, blessing abundantly all about them. He brings stones of beauty from the most unexpected quarter, and fits them into the temple with omniscient skill."

"I wonder if He sees any beauty in me," thought Olive, as the conversation closed, and she went to her room, saying over and over in her heart, "I want to be a 'living stone;' I want to be on the true foundation; I want to be numbered among the 'undefiled.'" That was a prayer in the eyes of Him who seeth every precious thing, and it was not unregarded.

CHAPTER VII.

SAPPHIRES.

" HAVE been reading over the description of Solomon's temple," said Ernest, the next afternoon. "It must have been very beautiful, but not nearly so large as I had been accustomed to think."

"It was not its size, but the great richness and elaborateness of its finish, which made it so highly regarded; so 'precious in the eyes of the people.'"

"I thought, before, too, that the carved 'lilies' and pomegranates were cut in stone. There were two hundred pomegranates on the two large pillars, which were named 'Jachin' and

'Boaz;' but they were of metal and wood. The pomegranates were of 'bright brass,' and the 'open flowers' were carved in fir wood, 'covered with gold, fitted upon the carved work.'"

"Did you notice the meaning of the names of these two pillars of the porch of the temple?"

"I do not remember," said Ernest.

"One, Jachin, means, 'He shall establish;' and 'Boaz' signifies, 'In it is strength.' They were both appropriate and beautiful mottoes for the entrance to a temple."

"The reason I read the account so carefully to-day," said Ernest, "was, to see if I could find where and how the precious stones were used; but I could not."

"We know they were there. Let us be as anxious to know where they 'are to come in,' in the spiritual temple. We can see them, as we look back, glinting here and there through the work of centuries; and there are more yet to be inwrought. Even as early as the second century, in the time of Polycarp and Martyr, they began to be seen; and they shine out more and more brightly through the mass of rubbish which continued to increase about the building, from century to century, for hundreds of years. One such gleaming jewel was set in its place in the year 295 after Christ, which sparkles with undimmed brightness. It was the sacrifice of worldly promotion and life itself, by a young man, for Christ. The beautiful spirit he manifested in his death gave such added luster, as makes it prominent in martyr story. He was but twenty-one. His friends were anxious to have him join the army of his country. As it was a pagan army, chiefly, he felt that he could not enter it. When examined, at last, by the proconsul, to see if his hight reached the required standard of a soldier, and be approved, he suddenly exclaimed, as the soldier's leaden badge was offered him, 'I shall take no such badge. I wear already the badge of

Christ, my God.' The reply of the heathen proconsul was, 'I will, then, instantly send you to your Christ.' 'Would you but do that,' was the reply, 'you would confer on me the highest honor.' Steadily refusing to have the badge placed upon his neck, he was sentenced to death. A new military coat had been prepared for him by his father, to be worn when he should enter the service. When he was led away to death, he requested that this coat be given to the soldier appointed to execute the sentence upon him."

"That was like Stephen, praying for his murderers," said Olive.

"It was the complete triumph of faith and love. His last words to the Christians standing near, and uttered with a countenance lighted with joy, were these: 'My dearest brethren, strive with all your power that the vision of the Lord may be vouchsafed to you, and that such a crown may be vouchsafed to you also.'"

"What was his name?" asked Ernest.

"Maximiliarius, of Teveste, in Numidia. I have never read this story without seeming to see the flashing light of a polished, precious stone, 'set for beauty' in Christ's own temple. As I said, such examples shone brighter from the surrounding darkness. Seeds of errors that afterward so entirely overran the church, marred its beauty and tarnished its glory, were already beginning to take root. But in spite of needless ceremonies, and errors of all sorts, and the cutting and hewing of men who worked with dead stones, God's eyes still kept fixed upon his building; He watched every stone as it was lifted to its place, and saw that they were so tried that they would endure any test the coming ages should bring. In the third century, when emperors set themselves to destroy the builders and their work, amid persecution, flames, and tortures of every kind, it still went on. Such men as Origen, who published the first Polyglot Bible, — or the Bible in different languages, — and Cyprian, of whom it has been said that 'he accomplished more in ten years than most men in a long life,' and whose writings breathed 'such a spirit of ardent piety that almost no one can read them without feeling his soul stirred within him,' — these men, and many others of that century, although the dust of the rubbish about them sometimes almost obscured their own vision, still kept nobly at work upon the wall, using the motto of Nehemiah's builders, 'The God of heaven, he will prosper us; therefore we his servants will arise and build.' "

"What did you mean by the dust getting into their eyes?" asked Mary.

Ernest smiled, and Mrs. Ewing replied, —

"Only that these good men themselves did not see the truth with entire clearness. They had true faith and Christian zeal, but not unmixed with errors. Origen's father was a living Christian, and he embraced the faith his father professed while very young; so that when his father was arrested (for he suffered martyrdom under the Emperor Septimius Severus), Origen himself longed to share his trials, even unto death. His mother prevented this by carefully watching over him, and keeping him out of the way of those who were on the alert to seize all who, like Origen, boldly proclaimed themselves to be Christians."

"Was he a martyr?" asked Mary.

"No; although he was only a little older than Ernest when he desired to share his father's fate, he lived to be nearly seventy years old; to be sure, he was persecuted, — exiled, imprisoned, and tortured, — and perhaps his sufferings hastened his death; but he lived to do great good for the church, notwithstanding the errors which crept into his teachings. He was a great student of the Bible, and expounded it to others wherever he had opportunity. He labored in Cæsarea and various parts of Palestine, and died two hundred fifty-three years after Christ, in the city of Tyre. His teachings were free, and in order to live he denied himself every luxury, and many of what we esteem the necessaries of life. He was placed in a tomb, 'near the high altar of the Cathedral of Tyre,' which was shown for hundreds of years afterward, but was lost, at last, in the destruction of that once magnificent city. Though the temples and palaces of Tyre have long lain in the dust, and its granite columns lie in the sea, the wonder to this day of the passing traveler, and almost the only tombstones of its ancient glory, the building on which Origen then worked, and of which he himself became a part, mocks the decay, by its still increasing strength and beauty. The persecutions of the Christians in the fourth century were still more severe. But in proportion to the wrath of those who opposed them, the life that was in the followers of Christ shone forth more and more brightly. Those who professed Christ had been hunted by thousands, and destroyed under Diocletian; one whole city in Phrygia, with all its inhabitants, having been burned to ashes because not an individual in it would offer sacrifice to the heathen idols."

"A whole city of martyrs!" said Olive.

"Yes; and the cruelties under another emperor, Galerius Maximianus, in this century, were too terrible for me to describe, as they are given to us, in glimpses, by historians. God himself took the matter in His own hand, and this emperor was attacked by a terrible disease, and his cruelties stopped by a miserable death. The martyrs of this century, so many in number, who 'came out of great tribulation,' with 'robes washed white in the blood of the Lamb,' have reminded me of sapphire stones, which are blue, and of great luster, hardness, and value; and some varieties of which '*become pure white* by exposure to heat.' Some species of sapphire were of a deep sky-blue, spotted with white, and spangled with stars of a gold color."

"It was the second stone in the breastplate," said Ernest. "I should like to see the starred sapphire."

"It makes me think," said Olive, "of 'those who turn many to righteousness;' for it is said that they 'shall shine as the stars for ever and ever.'"

CHAPTER VIII.

JASPER.

"CAN you tell me, Ernest," asked Mrs. Ewing, at the next 'hour,' "what great emperor lived in the fourth century?"

"Was it Constantine?"

"Yes; while other emperors were doing all they could to 'cause the work to cease,' he conceived it to be for his own interest to defend Christianity and protect Christians from further persecution. He conquered emperor after emperor in battle, until he remained, at length, sole ruler of the Roman empire, under the well-earned title of 'Constantine the Great.''

"Was he a Christian himself?" asked Ernest.

"This seems to have been more or less a matter of doubt among historians. It would be sad to think he had no place in a building which God enabled him so powerfully to protect. There are curious stories told of the motives which led him to favor Christianity, and also of his own conversion; but they are not wholly reliable. His father, while friendly to Christians, had also favored, at the same time, the pagan worship of the Romans. His mother, Helena, was called a Christian. Before his conversion, as it was called, Constantine was openly wicked. It is said that he had committed at least five murders among his immediate relatives (one of these victims being his wife, and another a son), and that, finding the pagan priests would not grant him pardon for sins so great as these, he turned to the Christian religion. We are told that when he inquired of one heathen philosopher what he could do to atone for his crimes, the reply was, that 'there was no remedy for such atrocious conduct as that of which he had been guilty.' A Christian bishop from Spain was then consulted, who told Constantine that 'in the Christian faith he could find a remedy for every sin;' and this assurance led him to seek it there."

"Is there not a story of his having had some wonderful vision of a cross in the sky?" asked Ernest.

"When making preparations for a battle with Maxentius, — so cruel and wicked a man as to be despised by pagans as well as Christians, — it is said that Constantine noticed that Maxentius looked for success to his heathen gods, and that this led him to think seriously where it would be best for him to turn for support. He at length determined to pray to the God of the Christians. Here comes in the story of the cross; that while he was thus praying the true God to reveal himself, he looked up and saw in the sky the form of a large, glittering cross, and above it the inscription, 'By this conquer.'

This vision was followed by a dream, the same night, in which Christ appeared to him with this same cross and motto, and told him to have a banner prepared after this pattern, and to carry it, as his talisman, into the battle. He caused to be made after the pattern he had seen, the resplendent banner of the cross,—called the Labarum,—on the shaft of which was affixed, with the symbol of the cross, the monogram of the name of Christ."

"What was that?" asked Olive.

"You have often seen it on books, I presume, and in various places: I. H. S.,—the first three letters of the Greek word Jesus."

"It must have been a beautiful flag," said Mary.

"I should have felt proud to march under it," said Olive.

"Would this dream and the result of it, if it made the emperor become a Christian, have been a miracle?" asked Olive.

"I should say, as much of a miracle as the

star that pointed the wise men to Christ," said Ernest.

"Yes, it would have seemed like a miracle. It has been doubted whether the story has much foundation in truth; and yet persons do sometimes now-a-days have wonderful dreams, and dreams which may make such an impression upon their minds as really to influence their actions. But this does not often occur, and there is no reliance whatever to be placed upon them. Whether Constantine did see the figure of a cross in the heavens or not, something, whether superstition, religion, or policy, led him to think a great deal of it, during his after life, as an emblem or symbol. And this, by some, has been considered one proof of the vision itself. There is still standing in Constantinople, that city of temples, a building of great magnificence and beauty, which was originally founded by Constantine, three hundred and twenty-five years after Christ, and whose ground-plan he had laid in the form of an immense cross, two hundred and sixty-nine feet long, and one hundred and forty-five wide. The hight of the building is one hundred and eighty feet. It was then called the Church of St. Sophia. It has undergone many changes, having been rebuilt in A. D. 538, by the Emperor Justinian; converted into a mosque, in 1453, by Mohammed II.; and partially built over, with some changes of architecture, as recently as 1847.

"Though built of brick, it is entirely marble-lined; and here, Ernest, you would find your favorite mosaic work, the ceiling and the arches between the columns of the building consisting of richly-inlaid patterns. There are between sixty and seventy columns upholding the gallery, some of which columns 'are of green jasper, and are said to have been taken from the celebrated Ephesian temple of Diana.'

"The symbol, as a foundation for so magnificent a building, strikes me as a very happy one, — the cross of Christ."

"It makes me think," said Olive, "of your text, 'Other foundation can no man lay.'"

"Was Constantine victorious in his battles after adopting the banner of the cross?"

"Yes; one historian, in speaking of this, says, 'He obtained the victory, and now felt that he was indebted for it to the God of the Christians.' The sign of the cross became his amulet; after the victory he caused to be erected in the Forum at Rome his own statue, holding in the right hand a standard, in the shape of a cross, with the following inscription beneath it: 'BY THIS SALUTARY SIGN, THE TRUE SYMBOL OF VALOR, I FREED YOUR CITY FROM THE YOKE OF THE TYRANT.'

"I have told you so much about Constantine because of the great changes he wrought in favor of Christianity in these dark days of persecution, and I can but feel that, through however much of error he groped his way into the light, he yet did find the true God, and did receive Christ as his tried and sure 'Corner Stone.' Certain it is, he guarded for others that building, whose foundation, the cross, is long enough and broad enough for the temple which shall fill the whole earth. He put an end to persecutions, and secured Christians against all forms of oppression, restoring the rights of which they had already been deprived, so far as lay in his power. In 314, Constantine became sole master of the empire of Rome. He began to suppress pagan worship from the first, but by degrees; as, occasionally, for some years after, he showed some respect himself to heathen rites. However, he at length publicly proclaimed his conviction that the only true God was the source of all his victories. I will read you one of these proclamations, which was issued by him to the eastern parts of the Roman empire: —

"'Thee, the Supreme God, I invoke. Be gracious to all Thy citizens of the eastern provinces, who have been worn down by long-continued distress, bestowing on them through me, Thy servant, salvation. And well may I ask this of Thee, Lord of the universe, holy God, for by the leading of Thy hand have I undertaken and accomplished salutary things. Everywhere, preceded by Thy sign, have I led on a victorious army. And if anywhere the public affairs demand it, I go against the enemy, following the same symbol of Thy power. For this reason, I have consecrated to Thee my soul, deeply imbued with love and with fear; for I sincerely love Thy name, I venerate Thy power, which Thou hast revealed to me by so many proofs, and by which Thou hast confirmed my faith.'"

"I should say that he was a pillar, — a jasper pillar, too," said Ernest.

"It certainly has the ring of a Christian proclamation. It was now his wish that all his subjects should embrace the same religion; still, he did not resort to force. 'Let no one molest his neighbor,' he said; 'the well-disposed must be convinced that they alone will live in holiness and purity, whom Thou Thyself dost call to find rest in Thy holy laws; . . . we have the *resplendent house of Thy truth,* which Thou hast given us in answer to the cravings of our nature.'"

"Had he yet openly professed to be a Christian?" asked Ernest.

"He had not yet been baptized. In fact, he was not ready to receive this outward symbol of his faith, so openly expressed in many ways, until the very last hours of his life.

"A singular mingling of truth and error seems to have followed him throughout his whole history. It has been said that 'even his statues' which he erected, 'halted between two opinions.' One of these statues was in the form of Apollo, but at its feet was buried a fragment of the true cross, and of the ancient Roman Palladium. The coins also which he had made bore on one side the name of Christ, and on the other the figure of a pagan god. Why did you speak of him as a jasper pillar, Ernest?"

"Because I thought that a very hard stone."

"It was this of which I was reminded by this very mingling of truth and error in his heart and life. The jasper is a precious stone, an opaque, crystalline substance, but debased, as it is called, with a mixture of earth. This accounts for its different colors of white, red, brown, and bluish-green. But, although debased by earths, it was regarded a very precious stone."

"Was it not the stone of which the walls of the New Jerusalem were built?" asked Olive.

"Yes; in Rev. 21:18 it says, 'And the building of the wall of it was of jasper,'" said Ernest.

"And because of his vacillation, at his death he was enrolled among the gods of the pagans. He was also made, by the Eastern church, a saint. But whatever may be said of these strange inconsistencies, he did a great deal, as all allow, for the cause of Christ. All his public acts, as emperor, were in favor of the Christian religion. He had been successful in battles, had become emperor of the world, and he now determined to assume the charge of the church. He was the first emperor who had assumed the power of sovereign in regard to the church, and it is said that he considered himself a 'bishop of bishops.' He wished to restore peace to the church, as he had to the empire."

"What troubles were there in the church?" asked Ernest.

"Many questions with regard to points of belief had occasioned strife among bishops and other Christians. Many of them were questions of no vital bearing, but upon which men spent much time in argument and foolish contention. The bishops had also suffered so much from pagan persecution, that they were rejoiced to have him assume this power. He called together the first general council of the church, — the great 'Council of Nice.'"

"When was that held?" asked Ernest.

"In the year 325 after Christ, and in the twentieth year of the reign of Constantine. It was customary for emperors in that day to celebrate every tenth anniversary of their reign. It is not known how Constantine celebrated the tenth anniversary, but, on the coming in of the twentieth, he assembled this great council of bishops and their attendants from all parts of the world, to adopt a settled form of belief, if possible, deciding with regard to certain questions which had long troubled the church. Men came from all parts of his empire, until two hundred and eighteen bishops had assembled; and, as each bishop was to be accompanied by two presbyters and three slaves, it was estimated that at least two thousand people were met at this time. Eusebius says, 'They came as fast as they could run, in almost a frenzy of excitement and enthusiasm.' It was a singular company. The young and the old, the ignorant and the learned, were there. They met in a large building in the centre of the town. The bishop of Alexandria, called the 'pope of Alexandria,' was there, — an aged man, with a shrill voice, but full of dignity. Arius was there, — called a 'heretic,' — whose peculiar belief had caused great discussions. He held that Christ was a created being, not equal to God, but first among created beings. His statements, said and sung, were such as these: 'God was not always Father; *once* he was not Father; afterwards he became Father.' Several of his followers were with him; and it is said that when he began to assert his belief, the bishops, who held him a heretic, 'raised their hands in horror, and kept their ears fast closed.' There were there, also, Copts, from Egypt, whose lives had been spent in deserts. Two bishops from Egypt were there, noted for the austere lives they had led. One, Potammon, from far up the Nile, had but one eye, the other having been taken out with a sword, and the socket seared by a hot iron. The other, Paphnutius, too, had lost an eye in the same manner, and was also lame from some former torture. Then there were bishops from Syria, some of them rude, wild men from the far East; others, scholars from Syrian cities. The bishop of Jerusalem was there, and, from a fortress on the Euphrates, a bishop named Paul, who, like the two African bishops, had suffered persecution, his hands being paralyzed from the torture of hot iron. Then there were four men, representing four orthodox churches, with their leader, a bishop of Nisibis, who had long lived a hermit's life,—in caves in winter, and in summer browsing on leaves and roots like a beast. He came dressed in a cloak of goat's hair. But all—the learned and the ignorant, those who had suffered torture, and lame, blind, halt, or withered—all were anxious and eager to get at the truth."

"I should think," said Ernest, "that it was very important that a man who should call and preside over such an assembly, should have been so decidedly a Christian as to leave no doubt of his character in that respect."

"That is very true; and we can not but think he was sincere in his devotion. When he entered this council, he was received with the greatest reverence. He was fond of dress, and having a person which was attractive and commanding, he impressed them with a feeling of awe as he entered the hall, wearing a diadem of pearls, and clothed in scarlet robes, which 'blazed with precious stones and gold embroidery.' The simple and the worldly, it is said, 'looked upon him as though he were an angel descended straight from heaven.' The 'Nicene Creed,' then adopted, has been almost universally received. If it will interest you to know what such a great and varied body of men agreed upon as the foundation of their faith, I will read you the creed they adopted."

"I would like to hear it," said Ernest.

"We believe in one God, the Father Almighty, Maker of all things, both visible and invisible;

"And in one Lord, Jesus Christ, the Son of God, begotten of the Father, God of God, Light of Light, very God of very God, begotten, not made, being of one substance with the Father, by whom all things were made, both things in heaven and things in earth, who for us men, and for our salvation, came down and was made flesh, and was made man, suffered, and rose again on the third day; went up into the heavens, and is to come again to judge the quick and dead;

"And in the Holy Ghost.

"But those that say, 'there was when He was not,' and 'before He was begotten He was not,' and that 'He came into existence from what was not,' or who profess that the Son of God is of a different 'person' or 'substance,' or that He is created, or changeable, or variable, are anathematized by the Catholic church."

"I was going to say, before you read that last part," said Ernest, "that I should have thought Arius would have 'held up his hands in horror, and closed his ears and eyes.'"

"Did Constantine like the creed?" asked Olive.

"Yes, the whole of it. He seemed to revere the bishops, and regard their decision as inspired. He denounced Arius and his followers, ordering his books to be burned, and the penalty of death to be inflicted upon any who should read them. This was the first written creed, and was subscribed to in written form, many of these names remaining till the next century. A bishop of Spain, Hosius, signed it first, in this way: 'So I believe, as above written;' and the presbyters of the bishop of Rome signed it for him as follows: 'We have subscribed for our bishop, who is the bishop of Rome. So he believes, as above written.'

"Some time after this, Constantine determined to make Byzantium, now Constantinople, the seat of his empire. And here he spent the last years of his life. At the age of nearly eighty, his mother, Helena, made a pilgrimage to Jerusalem in search of the Holy Sepulcher, and of the very cross upon which Christ was crucified. It was said that the sepulcher had been revealed in a vision, and the three crosses also, which were found buried near it."

"How would they know Christ's from the ones on which the thieves hung?" asked Olive.

"It was said that those who touched one of these three crosses were healed, and that one dead body was restored to life by its contact. So, of course, that decided it to be the Holy Cross!"

"Did Helena find it?" asked Olive.

"Yes; and sent a part of it to Constantine, which, as I told you, he had buried beneath, or enclosed in, one of his statues. The greater portion of this cross was kept at Jerusalem, and set in a case of silver. Multitudes of pilgrims, through a long series of years, visited Jerusalem, and obtained a small fragment of this 'true cross.' So that St. Cyril remarked that 'the whole earth was filled with this sacred wood.'"

"He, then, did not believe it was the true cross?" said Ernest.

"Oh, yes; but he accounted for this by a miracle, which he compared to the miraculous multiplying of the loaves and fishes by Christ. Some of the monkish writers have insisted that, in spite of all the fragments taken from it as relics, it still remained undiminished in size until the year 613, when Jerusalem was taken by the Persians, who carried the cross home with them, and kept it fourteen years, when it was again restored to its former place on Mount Calvary, by the Emperor Heraclius. It is said that this emperor, laying aside his crown and robe, carried the cross on his own shoulders to the sepulcher, and appointed an officer to its special care. The anniversary of this event, the 15th of September, is still kept in the Greek church among their festivals, and called the 'Exaltation of the Cross.'"

"What became of it, after all?"

"It has had quite a romantic history. And no one knows where the truth 'comes in,' as Ernest says. It is said to have been taken from Jerusalem by the Arabs, a party of whom invaded Palestine, and took the cross to Constantinople only eight years after Heraclius secured it, in the year 636. Here, curiously enough, it was kept in the Church of St. Sophia, founded, you remember, by Constantine; and the honors here paid to it are described by Bede, an English historian. It is said that on the three solemn festivals of the year, 'its costly case was unclosed; when a grateful odor pervaded the whole church, and a fluid resembling oil distilled from the knots in the wood, the least drop of which was thought sufficient to cure the most inveterate disease.'"

"I should think the people could *see* and *know* whether this was so," said Olive.

"It is strange how much people can be made to believe, with their eyes and ears open, and bearing no witness to its truth."

"What became of the cross at last?" asked Mary, again.

"It is said to have been taken and re-taken by various persons until 1238, when it was purchased by Louis IX., king of France, and placed in a chapel which he had built for it, and where it remained for three hundred years, when it suddenly disappeared, and no trace was again found of it. Some of the king's subjects accused him of having secretly sold it to the Venetians, and, to appease their wrath at this suspicion, he had a new cross made of the same shape, size, and appearance, and which, he said, was but little inferior, as an object of worship, to the real cross. And this satisfied them! I can imagine Christ Himself looking down from heaven upon these deluded wood-worshipers with infinite pity, and exclaiming, 'Oh that they were wise! that they understood this!'"

"I do not suppose there was any harm," said Ernest, "in people desiring a bit of the true cross, if they were perfectly certain it was a part of it; I should value it myself, but not as anything to be worshiped."

"There was the danger. They took the wood for the foundation, instead of the *Stone* which God had chosen, and tried, and laid Himself. They clung to material relics, and failed to see where their true hopes should be fixed; just as the Jews worshiped the brazen serpent, for hundreds of years after its work was done, making a god of the brass, and calling it 'Nehushtan.'

"This passion for relics in Constantine was one of the *earth stripes of the jasper*," said Mrs. Ewing. "At the same time that he received a part of the Holy Cross, he also had sent him nails, which, it was pretended, held Christ upon it."

CHAPTER IX.

A STUMBLING-STONE.

"WAS the next emperor as favorable to Christians as Constantine?" asked Olive.

"His three sons succeeded him in the empire, and, while they favored the Christian religion, as he had done, they were not equally lenient toward the pagans. They ordered all heathen temples to be destroyed, and death was made the penalty for consulting pagan oracles."

"What was the name of these three brothers?" asked Olive.

"Constantius, Constantine, and Constans. But they had wars between themselves, and Constantine was killed in one of these. Constans died in 350, and Constantius was left, as his father had been, 'sole master of the entire Roman empire.' He still made it death to use pagan rites and ceremonies."

"What sort of martyrs would those be?" asked Ernest; "those who suffered death, I mean, because of a *false* religion?"

"That is only another proof that *suffering* does not work *salvation*. They could not earn the title of 'undefiled' by submitting to death on account of their false faith. Constantius was the means of the destruction of many of the pagan temples, and was so severe in many ways with regard to pagan usages, that even Christians felt that he injured their cause; and in the end it proved to be so. It was almost certain that, if a pagan emperor should succeed Constantius, all this severity would be returned upon the Christians.

"His own nephew, Julian, was his successor. He had been carefully and severely trained in the Christian religion. He was well taught in the Scriptures, but, as there was much of outward constraint in this, when he became free to choose for himself, a reaction took place, which led him to turn to paganism, and do all in his power to promote it. He not only took part with pagans, but also with the Jews, against Christians.

"Their temple — which had been destroyed, 'so that not one stone was left upon another,' as Christ had predicted — he encouraged the Jews to rebuild. But this was not to be done. The Lord had consigned it to utter destruction because of their sin. Olive, you may read a passage which shows this, in Micah 3 : 12."

Olive turned to the passage, and read, "Therefore shall Zion for your sake be plowed as a field, and Jerusalem become heaps, and the mountain of the house as the high places of the forest."

"The Jews could not believe this ruin was to be permanent. So, with Julian's permission, they began to lay a foundation for their temple. But the Lord was against them, and 'except the Lord build the house, they labor in vain who build it.' Their work was stopped, suddenly and effectually, by balls of fire issuing from the ground, and by earthquakes, so that they could not go on. This has been doubted by some; but its truth seems to have been very fully proved. And so God stopped the false building of the Emperor Julian, who would have crushed out Christianity, and erected, on a false foundation, a temple in honor of unknown gods. He taught the people that they were to worship not only groves, and altars, and images, but even the temples themselves. He also taught priest-worship, even of unworthy men, saying, 'So long as he sacrifices for us, and stands before the gods as our representative, we are bound to look upon him with reverence and awe, as an organ of the gods most worthy of all honor. . . . In performing the functions of his office, he should also wear the most costly apparel.' You see here how errors crept in from century to century. Julian had his priests clothed in purple, and decorated with garlands. He had pagan temples rebuilt, and the gods returned to them. He laid wily plans to oblige Christians to worship, outwardly at least, the gods of the pagans. His own statues, set up in open places, had images of Jupiter, and Mars, and Mercury, connected with them. 'A Jupiter over his head reached down to him the purple mantle and crown, while Mercury and Mars looked on with an approving smile.' As it was customary to bow before his statues, of course, in doing this, in one sense they would bow to the gods."

"That would not have been worship of the gods, though, would it?" asked Mary.

"Not really; but it was a mortifying thing for them to do. He afterwards went to Antioch, a city where were a great many Christians, and tried to force pagan rites upon them. He restored there the 'feast of Apollo,' hoping the people of the city would join in the display, and was much chagrined because they showed contempt for the festival. And, because 'not an individual brought oil, to kindle a lamp to the god, not one brought incense, not one a libation or a sacrifice,' and 'but one solitary priest appeared, bringing a goose for an offering,' he was indignant and angry. Before leaving the city, he placed over them a cruel and tyrannical governor, directed some to be put to torture for crimes with which he chose to charge them, and ordered their Christian temple to be closed. *He* was a stumbling-stone in the way of the church. But as balls of fire prevented the Jews building against the will of God, so he was not permitted to work longer against the church of God. He died in battle in the year 363, and, as one of his historians remarks, 'At a single blow, the frail fabric erected by mere human will was dissolved.' Christ became to him a 'stumbling-stone.'"

"Other foundation can no man lay," said Olive.

"You may look through all history, from that day to this, and you will find the same thing. Error and falsehood may beat against the walls of the true temple, but their wrath is in vain; they may try to build in opposition to it, but their work will soon be confounded. It was through such oppression and bitter opposition that Christ's building, 'fitly framed together,' slowly went on."

"It seems as if there is nothing in our time to hinder its going forward," said Ernest.

"There are but few countries now where persecution prevents Christian workers from building. And yet there have been martyrs in our own times. Some of our own missionaries, such as Williams, and Coffing, and Merriam, have fallen victims to heathen cruelties, as did the early Christians; but their work is not lost. There are, however, enemies always ready to annoy and hinder, and, if possible, prevent this Christian building from going on undisturbed. We have enemies very near us, subtle and strong, who often attack us in our work, and well nigh discourage us. There are giants in the way. Unbelief whispers, 'You are not fit for such a temple. You are a worthless, untried stone. The Master Builder would scorn such material.' Or, if not attacked ourselves, we are working upon some stone in the great quarry, and the giant Sloth comes in, and says, 'It is not worth the trouble you are taking. It is too far away from the temple. If you should get it hewn and polished, you could never get it lifted into place.' Or Selfishness, another giant, whispers, 'I would spend more time on my own marble, and polish it with superior care; let the other blocks lie as they may; what is that to you?' Or the giant Envy comes along, and seeing your work carefully done, and about to become fit for a palace, defaces it with a blow here, and a rap there, until you grow weary in the effort to become fitted for a place in the great building, at all."

"I never thought of those enemies," said Olive.

"That was what Paul meant when he said, 'When I would do good, evil is present with me.' But there is the promise to 'him who overcometh;' and we fight these giants with prayer, as we work, hoping 'the pillar' will one day appear."

"Oh, blest is he to whom is given
The instinct that can tell
That God is on the field, when He
Is most invisible."

CHAPTER X.

HIDDEN STONES.

"IN the fifth century, the darkness and superstition which closed about the church of Christ were greater than during the reign of Constantine and Julian. Even Christians failed to separate the rubbish from the stones, and builded poorly, and with much material that would prove but dross, when 'tried as by fire.' Even the bones of the martyrs were thought to have power, by being touched, to heal diseases. Images of saints were worshiped. Pilgrimages were made to places esteemed holy, some of them almost too ludicrous to mention. Images were placed in gorgeous temples. Relics were valued as sacred, and silver boxes and altars made to hold them. Private confession to a priest first was allowed in this century. Bodily torture and penance were undergone with a view of pleasing God. Some zealous Christians went about almost without clothing, making a diet of grass and hay, living alone or among beasts, standing on the tops of pillars (such were called 'pillar saints,' not in the sense of those who 'overcome' in God's way), pining away in dens and caves; and such were accounted saints!

"But in the midst of all this, God's eye was upon those who looked to him in the simplicity of a pure faith. Vigilantius was one of these. He taught boldly that these superstitions were foolish and harmful; that many of them were borrowed from the pagans, and were all displeasing to God.

"In the sixth century, through the same groping in darkness, the work went slowly on.

The darkness was greater, the superstitions more numerous, and those who still wrought on the temple fewer; but the work did not stop. It seemed as if Satan was indeed leagued against it, and the names of the true workers are so few that we can scarcely find them. There were in England converts to Christianity through the influence of missionaries sent them under one Augustine by Gregory the Great. Among them was Ethelbert, king of Kent, who did what he could for the religion of Christ, changing pagan temples into Christian churches, and outwardly, at least, helping build for Christ.

"Again, in the seventh century, it seemed as if Satan's enmity to the church was about to end in its defeat and destruction. In 612 Mohammed appeared, and laid his foundation for a great opposing work. He drew many, who had appeared almost fitted for the true temple, from their place, and rejoiced to see them on the foundation he had attempted to lay. It has been said that, 'during this century, true Religion lay buried under a mass of senseless superstitions, and was unable to raise her head.' It is a sad picture, but there were some who could still say, 'God is with us,' and who 'forsook not the walls of Jerusalem.'"

"I wonder that the good did not entirely die out of the world," said Olive, who seemed oppressed by the darkness, as if it might still be felt.

"And why did it not?" asked Mrs. Ewing.

"I suppose because the work was of God."

"That is just it. We can not tell why He has so great patience with the wicked; but we do know that the time will come when all their rage shall cease; when they shall receive the reward of their doings, and there shall be 'none left to hurt or destroy in all God's holy mountain.'"

"And there are opposers among us now?" said Olive.

"Yes. All who do not work for Christ are

against him. It is a fearful thought. It is easy for us to judge others, but hard to see that we ourselves may be disturbers and hinderers in this glorious work. With all the light we have, there is less excuse for us than for those early opposers of the religion of Christ."

CHAPTER XI.

BRECCIATED MARBLES.

"WHAT have you there, Ernest?" asked Mrs. Ewing, as the children again met for their afternoon talk.

"Some specimens of marble," said Ernest. "Mr. Sterling gave them to me. He brought them from a marble shop, and as I was specially interested in two or three kinds, he gave me these. I did not know that there were so many kinds of marble, or that it was found in so many different colors. This is very beautiful," said he, holding up a small piece of a fine rose tint, mottled with white. "It came from Vermont. I have never seen much but white marble."

"Do you remember that marble factory, or shop, where we once went to see them sawing the slabs for building?"

"I do," said Mary. "How slow the great saw went through the stone!"

"Those were not very large or thick stones. They were to be used for facing brick walls, — a sort of veneering in stone, as rose-wood or mahogany is put in thin layers over furniture, or fancy and ornamental boxes. This piece of marble," said Mrs. Ewing, taking up a curious stone, polished on one side, "they would find difficult to saw, even in thin slabs for facing. It is called 'brecciated' marble, — one of Nature's composition stones. If you look at it closely, you will see that it looks as if made up of pieces of different sorts of stone, — and this is, no doubt, often so, — but so thoroughly cemented by other earths as to seem one solid stone. There are several kinds of this brecciated marble, some of which are very curious, and never seen, perhaps, in this country, except as rare specimens in the cabinets of geologists. One of these varieties is of a black ground mottled with spots, and called either 'the full mourning' or 'the half mourning.' The breccia marble of red, spotted with white, from Italy, is more common, and has been used in ornamental parts of churches in this country. The breccia marbles of this country are found in all colors, — drab, red, salmon, and white, — in different styles of mottling, shading, and blending."

"Why would it be harder to saw this kind of marble?" asked Olive.

"Because, being composed, as I said, of different kinds of stone, these kinds vary in hardness, so that working it is a slow process. This is especially true of some varieties."

"It seems to me that some people resemble that kind of marble," said Ernest. "They seem so different at different times, — 'good in spots,' as Fred Green says."

"Some Christians might seem a little like it," replied Mrs. Ewing.

"I did n't say Christians," replied Ernest.

"But do n't you see, on the polished side of this breccia stone, the places show where the fragments join almost as plainly as on the rough side? You have, no doubt, heard persons say that the scars of sin remain even when they are forgiven, as the scar of a wound remains when the wound is healed."

"It would make one wish to be careful how one lived," said Olive. "But I think this variety, especially where it is polished, adds to its beauty."

"Is there any sense in which this might be true of those who had sinned much?"

"I do not see how there can be," replied Olive.

"Nor I, unless the beauty that love throws over a person's character adds to it enough for this. You know it is said, 'He who is forgiven much loveth much, but unto whom little is forgiven, the same loveth little.'"

Olive smiled as she caught her mother's meaning, but said, "I should want the love without the scars."

"Mr. Sterling showed me another curious specimen of marble, which I almost coveted, — a little corner of it, at least. And it might also remind one of scars, or something of the sort; it was marble, full of little petrified shells. This was a sawed and polished specimen, and the shells were cut across so as to look very curiously, — a little like the inside of your glass paper-weight."

"I have seen this 'Kilkenny marble,' as it is called."

"I should not like to have faults preserved in that way," said Olive; "petrified in."

"No faults will remain to mar the beauty of our characters, or prevent our becoming perfect building stones, if we are fitted for the true foundation-stone by the Master Builder. What remains, after we have been finally tried, shall be 'without spot or wrinkle, or any such thing.' The errors of the Christians, in those dark days, would melt away, one after another, beneath the tests applied to them, and all who had any genuine faith would be able to stand. But their peculiarities of character might remain without marring their beauty. This will doubtless be so."

"What century was it," asked Olive, "when you said Religion seemed buried in superstition, and unable to raise her head?"

"The seventh. But, as I said, there were some even then guarding the wall, and even bringing in, now and then, freshly-hewn stones for the temple. I told you of missionaries sent to England, and of the conversions among the people there. 'You know they are to come 'from the north and from the south, from the east and from the west.' We may work in any quarry, if it seem ever so distant, with the certainty that no living stone will ever fail to be brought to its fitting place. There is a building in South Carolina, every block of which was quarried, and exactly fitted for its place, in New England. Each stone was marked with a number corresponding to its place in the plan of the edifice. God numbers every stone in the world's great quarry, and its corresponding place will surely be met. 'Look unto me, and be ye saved, all ye ends of the earth.' 'Fear not, for I am with thee; . . . bring my sons from far, and my daughters from the ends of the earth, every one that is called by my name; for I have created him for my glory; I have formed him, yea, I have made him.' This has been the encouragement of God's people in the darkest days of the world's history. In the eighth century, belief in miracles of the most foolish kind was added to other superstitions, until even the best of people found it difficult to get at simple gospel truth. Of one of the saints of that time, who owned a mill, it was said that, while he stopped to say his prayers, the mill would go of itself."

"That might be," said Ernest, "if the prayers

were very short, and the mill in full force when he began."

"The explanation is as good as the miracle, at any rate. It was also believed that a person who peered through a crack in this mill, to see how it was possible such a thing could be, was struck with blindness for so wicked an act."

"That reminds me," said Olive, "of the people in India, who believe Juggernaut must be represented in an unfinished condition, because a man looked through a crack and saw the maker of their first Juggernaut before he had completed the idol."

"Of another saint it was said that his presence could make a child's cradle rock, untouched by any person; while, if one should attempt to rock it, it would be impossible to move it."

"I suppose 'spirit-rappings' began in the eighth century, then," said Ernest.

"It looks more like witchcraft than the work of a saint. Many of their stories were more ludicrous than this; and some of their superstitions vastly more troublesome. Evil spirits were believed to come and take saints bodily and carry them into full view of the home of the lost, where saints, after they had been sufficiently frightened with the horrors of hell, would come to their rescue, and bear them to their homes again. Even among the priests and monks, — who were supposed at this time to have the greatest amount of learning, — the ignorance was so great that they scarcely knew more than the Lord's Prayer, the Creed, and the Psalter, — which last was a collection of one hundred and fifty devout sentences. There was also, in this century, much of image-worship; so that, when an attempt was made to abolish it, a war of several years' duration was the result.

"In the year 754, a council of three hundred and thirty-eight bishops met in Constantinople to settle the question, and the result was the decision that 'all worship of images was contrary to the Scriptures, and to the sense of the church in its purest ages; that it was idolatry, and forbidden by the second commandment.'"

"I should think a child might decide that much," said Olive, "without three hundred and more bishops!"

"One would think so. This was a Grecian council. The Romans did not like their decision, and they called a council about thirty years later, and decided that 'all who should hold that God only was to be worshiped should be punished.' The English, French, and Germans did not wholly agree with the Romans. They wished to keep the images in the churches, but not offer them worship. So they had a council at Frankfort, in Germany, under Charlemagne, the greatest emperor of the time; and here it was decided to abolish image-worship. Although it was the Greek, or Eastern church, as it is called, they afterward, in 842, decided it to be right that 'the images of Christ, of the Virgin Mary, and of angels and saints, should be worshiped.' But finally they took the position which the Greek church still hold, — that, while carved, or molten, or sculptured images should not be revered, pictures, which they say are not images, are worthy of adoration!

"The worship of relics still continued. Superstitions with regard to the Lord's Supper were singular and annoying in the extreme. Severe penances were connected with any impropriety in its performance. The minister who should spill a drop of the wine must undergo a three days' penance, after sucking the drop from the altar-cloth where it fell. If any one dropped the bread, or let fall the cup, he must, by way of penance, sing fifty psalms. If the wine was not well preserved, so that its flavor or color was not perfect, penance must be undergone for this. A penance of forty days' duration was exacted if the cup should be turned up, instead of set down, at the close of the service."

"It seems like child's play," said Ernest.

"It was little better. Everything, with most of the people, was outside worship. Pilgrimages, penances, relic-worship, the founding of churches and chapels as a means of appeasing the wrath of God for sin, — these were the exponents of the thick darkness through which God allowed some rays of light to gleam. One of the great builders, himself a pillar of the church at this time, was Bede. He was called the Teacher of England. He wrote extensive commentaries on the Bible, and also the Church History of the English nation, from the invasion of Julius Cæsar to his own time, — a book which received the honor of being translated into the Saxon language by King Alfred. He educated many teachers who, like him, bestowed great pains upon the study of the Bible. His death among his pupils, — engaged until the last moment in the translation of the Gospel of John, — was one of the most beautiful on record. And you will be glad to know how his influence lived after him. Alcuin, a successor of his in the great work of teaching the Scriptures, was born the same year in which Bede died. Bede's mantle fell upon him. He was educated in a school in York, England, whose superintendent, Egbert, was a friend and pupil of Bede. He took the torch from his dying master, and its light burned brighter and brighter, until hundreds of other torches were set ablaze by its flame. Alcuin became the head of this school after Egbert, and was known as 'the great teacher of the time.' Scholars flocked to him from distant places, until his school became a center of great influence. He was invited by the Emperor Charlemagne to Germany to become a teacher and guide to his people. Charlemagne's rule extended over many countries — France, Germany, Italy, Spain, and many Saxon tribes. He exerted great influence over all his dominions, having them visited often by delegates, and keeping a careful oversight of the whole himself. While he did not neglect trade and industry, and the physical welfare and improvement of his subjects, he bestowed special care upon their instruction in classical and all other useful learning. He invited learned men from all parts of the world to his court, to become teachers in the different branches of science in which they excelled. Some were to teach in astronomy, — of which he was so fond himself as frequently to rise in the night and gladly give up his sleeping hours to the study of the stars, — some to teach history, some rhetoric and logic, and others to teach theology. In this way Alcuin was called to his court. And it has been said of this York teacher, that he was accounted 'the leading spirit of this aggregation of teachers.' He originated a school in which men were thoroughly taught, and fitted to be sent out as teachers into other provinces, and which formed a society of learned men, of which Charlemagne himself and several of his immediate family were members.

"The ignorance of the bishops, of which I have been telling you, struck Charlemagne, as revealed in the letters they frequently wrote him, petitioning him for some favor which they desired. He issued a circular, recommending them to be more diligent in scientific studies, as a means of increasing their ability to understand and explain the Holy Scriptures. Alcuin had been sent from York, while yet a teacher there, on a mission to Rome, by the archbishop of York. On his return he was met at Parma, in Italy, by Charlemagne, and urged to remain with him, and take charge of these institutions that he was about to set up. Having returned to York, and got leave both of his king and the archbishop, he went back to Charlemagne, who granted him the position of teacher of the monks in two monasteries, and especially the superintendence of a school near his palace for young men of noble parentage, called the Palatine School. He was the friend, instructor, and the consoler under affliction of the emperor himself, and when absent from the emperor, he wrote to him freely, keeping up the correspondence during his life. Alcuin, like Egbert and Bede, gave great attention to the study of the Scriptures, and his influence for good at such a time, in his high position, was immense. The emperor called him 'his most beloved teacher in Christ.' For eight years he labored in this manner for the emperor, and then went back, spending two years in York, when he returned, and again taught under Charlemagne, working with great zeal and success until old age caused him to seek a respite from such arduous cares in greater retirement. He was given the charge of instructing the monks in the Abbey of St. Martino. Here he labored as long as his failing strength would allow, when he was permitted to choose from his own scholars those who should fill his place. Like the good Archbishop Leighton, whose wish with regard to the place of his death was granted, so was it with Alcuin."

"How was that?" asked Olive.

"Leighton often said he would be glad to die at an inn. It seemed a singular wish, but he did die, while upon a journey, at an inn. Alcuin wished to die on the return of the festival of Pentecost, and he passed away on that day,—May 19, 804. So, you see, even in the darkness of the most superstitious ages, God fulfilled His promise to watch over His church. You may read one of these beautiful promises, Olive, in Isaiah 61 :. 67."

Olive read,—

"'I have set watchmen upon thy walls, O Jerusalem, which shall never hold their peace, day nor night. Ye that make mention of the Lord, keep not silence; and give Him no rest, till He establish, and till He make Jerusalem a praise in the earth.'"

"And although the best of these men were tinged with the errors, and perhaps a little marred by the superstitions, of the day, like the breccia marbles they bore the polish of the Master, and stood before the world with added beauty."

CHAPTER XII.

ONYX STONES.

"WHO was the successor of Charlemagne?" asked Ernest, the next Sabbath afternoon.

"His son, who was called 'Lewis the Meek.' He too, like his father, favored Christianity, and did much for the upbuilding of Christ's true kingdom. He sent religious men into Sweden and Denmark, and gathered material for this great living temple from those foreign countries.

"The emperor of the Greeks, Basil, of the ninth century, sent Christian men to Russia — the first missionary work ever attempted in that country. Many were influenced to accept the truths taught them, and although much of the rubbish of superstition remained in their best teachings, there was enough spirit and life to impart life to others. But there was so much of persecution, opposition, plunder, and robbery from other nations, that Christians were very thoroughly tried, and many who had called themselves such gave up the profession, and, for the sake of quiet and security, embraced the false faith of their conquerors. A question was agitated in the ninth century, somewhat like the one which has been lately troubling the great council at Rome."

"This question is the infallibility of the pope?" said Ernest.

"Yes. There was a great deal of vice among the Romanists in that day, but their influence was considerable. They began in this century to put forward the claim, that the 'bishop of Rome was constituted by Jesus Christ a legislator and judge over the whole church, and, therefore, that other bishops derived all their authority solely from him, and that councils could decide nothing without his direction and approbation.' In order to prove this claim, they must have something more to show than assertion merely. Ancient writings were forged by the monks. 'Decrees of councils, never before heard of, were now discovered, by which the universal supremacy of the pope was established from the earliest times.' These new builders attacked the workers on the one only foundation with more bitterness than was displayed by Sanballat and Tobiah, and the other enemies of the Jews in Nehemiah's day, when they 'conspired all of them together to come and fight against Jerusalem, and to hinder it.' And there were many who were hindered, and said, as Judah did with regard to those enemies, 'The strength of the bearers of burdens is decayed, and there is much rubbish; so that we are not able to build the wall.'

"But there were others who said, 'Be not ye

afraid of them; remember the Lord, which is great and terrible;' and these builded, 'every one with his sword girded by his side.' 'They which builded on the wall, and they that bare burdens, with those that laded, every one with one of his hands wrought in the work, and with the other hand held a weapon.' 'And half of them held the spears, from the rising of the morning till the stars appeared.' That was the way Christians worked in those days, and that is the way they always must work; for there are always enemies ready to hinder, whether they attack suddenly and in troops, or slowly, and one by one. So that the watchword which has served the church in all ages remains still the best for us, and it was given by Christ Himself: 'What I say unto you, I say unto all, Watch.'

"King Alfred, in England, who translated Bede's Church History, did much in this century for the truth. He was a friend to learning, and it is said by some that he founded the university of Oxford. Many attempts had been made to destroy the Scriptures, but copies of them had always been preserved, even in times of greatest peril. In this century, so much of superstition prevailed as to affect even the best of those who upheld Christianity. It would be as difficult to decide whether he himself were a Christian, as whether Constantine was; and we had better give them the benefit of the doubt. For the same reason that Constantine reminded us of jasper, Alfred may be represented by the onyx, which is strongly contrasted in its different colors, and one layer will be found clear and transparent, joined to another dark and opaque. He attempted to civilize and Christianize those whom he conquered in his various battles. He was a superior scholar and writer. Like Constantine, he, too, was surnamed 'The Great,' and seems equally well to have earned the title.

"In many countries, in this century, image-worship was continued. 'Candles were burned before images, incense burned to them, supplications made before them; colors were scraped from pictures, and mixed with the wine to be drank at sacraments; and the bread was placed in the hands of images, so as to be received as from them.' "

"I should think *that* more a lack of common sense than ignorance," said Ernest.

"And even this was by command of the popes, who established image-worship throughout the West. In the tenth century there was still greater darkness; yet the church was not defeated, and continued slowly to increase. The Nestorians, among whom we now have missionaries, had then the true faith, and sent missionaries themselves to foreign countries, even to the borders of China. Christianity was embraced also in Poland and Hungary, and extended in Russia. Pagan nations still troubled the Christians by persecutions, trying to tear up the very foundation stones of their building; and the tenth century was called, from all these growing evils, the 'dark age,' the 'iron age,' and the 'leaden age.' The priests and bishops were worldly, sensual, and even criminal, fond of traffic and pleasures of every sort. It is said of one of them, who was particularly fond of horses and hounds, that 'he kept two thousand horses, which he fed on nuts, and fruits, and odorous wine'! These bishops were called, in this 'iron age,' 'bishops of the world,' instead of bishops of Rome; and some were ready to admit that 'bishops receive all their power from God, but only through St. Peter.' In the eleventh century, the Crusades commenced."

"What were they?" asked Olive.

"Something about taking Palestine from the Mohammedans," said Ernest. "I've read of Peter the Hermit, who led the first army, many of whom perished. They carried banners, with the symbol of the cross, and wore a cross made of some bright-colored material, on their shoulders. But I do not know whether much good was accomplished by them."

"The soldiers were many of them men such as we should not call 'soldiers of the cross,' making havoc and destruction through the countries as they marched; but they at last conquered the Turks, and took possession of Jerusalem. There were still in this century true and fearless builders, who taught the truth in spite of persecution, and became exiles, if not martyrs to their belief.

"The pope, Gregory VII., proclaimed, in this century, the perfection of the Romish church. One of a series of propositions that he put forth at this time, was 'that the Romish church never erred, nor will it, according to the Scriptures, ever err.'"

"Did he give the proof from the Scriptures?" asked Olive.

"The proofs did not appear with the assertions! Among other things, he assumed the power 'to depose emperors,' and declared that any sentence he pronounced could be recalled by no one, but '*he* could review the decisions *of all others.*' The Roman King, Henry IV., was obliged to humble himself before this pope, and was compelled, while waiting on him for the pardon of his sins, to stand, in midwinter, three days, barefooted and bareheaded, and in a miserable dress, within the castle walls, making confessions, before he received absolution."

"That is very different from Christ, waiting at the door of our hearts," said Olive, "for us to let him come in with our pardon."

"As different as the spirit of the two religions throughout. But there was much of compulsion used among Christians even, in their efforts to extend Christianity. Some good may have resulted from the wars carried on to force men to receive a new belief. But God says it is 'not by might, nor by power, but by His Spirit,' that the work is to be done.

"In the twelfth century the Mohammedans caused the Christians great trouble, waging constant wars with them, so that, fearing they should lose the hold they had gained upon Palestine by the first crusade, they undertook a second, instigated by St. Bernard. The King of France aided them, as well as the Emperor of the Germans, and their armies started for Palestine in 1147; but they were defeated, perishing by the way in great numbers, and so divided among themselves that, when their reduced and broken armies met in Palestine, they could do nothing but march back over their weary route, having lost in the expedition nearly two hundred thousand men.

"Bernard's life was a very eventful one. Onyx-like, it was full of contrasting layers of light and darkness. He was a Frenchman of noble descent, early trained for the church, and desirous of leading a monk's life. He fasted, and used many methods of mortifying the flesh, until, worn and haggard, he became a 'fearful witness of the struggle of the soul in its contest with the body.' He was chosen abbot by twelve brother monks, and they formed themselves into a community, living in cells which they built in a wild, robber-infested mountain gorge, named 'The Valley of Wormwood,' where many visited them, and where, it was believed, miracles were performed by Bernard upon the sick and infirm who came to him. He was saved from self-destruction in this way of living by a friend, restored to health, and induced to live a life of less austerity. After living twelve years in a monastery, he became a great teacher, preacher and writer, and, after the failure of the crusade, he battled with various forms of error at home, and wore himself out in what he believed true work for the church. He died at the age of sixty-three, in the year 1153, and was afterward placed among the saints. It was said of him, 'Few have loved the church with more steadfast and unselfish devotion.' Few have rendered it more signal services. The church knew him as a trusty servant, faithful to his profession, terrible to all its foes. He wrote hundreds of sermons, and crowds were moved by his preaching. What he did that shall stand as a true builder, will be revealed in the great test day."

"I thought he was a Catholic," said Ernest, "and was made a saint."

"Neither the one nor the other will hurt him, if his foundation-stone was Christ. If he built upon that, the name he was called by while a worker, the name he received after death, will neither hinder nor help in the matter of his salvation. The foundation is broad and sure: the one thing is to be upon it. It would be sad for such a worker to find in the end that his life had been a mistake,—that he had hewn and polished a stone which God had not numbered for His temple, and that he had prepared many blocks for a building where they should find no place.

"It is of the first importance to all to be sure that they have truth on their side. The world is full of workers. The great question is, How much of the work shall stand? Every one whose trust has been placed in Christ is safe. He says to all such, 'I have covered thee in the shadow of mine hand, that I may plant the heavens, and lay the foundations of the earth, and say unto Zion, Thou art my people.'"

CHAPTER XIII.

AGATES.

"WAS there not a third crusade?" asked Olive.

"Yes; but it, too, was unsuccessful. Frederick I., of Germany, began this crusade in 1189. He was himself drowned in Syria the next year, and most of his army returned. The same year in which Frederick was drowned, the kings of France and England — Philip Augustus and Richard the Lion-hearted — undertook the same work, and reached Palestine, but after a partial success, returned, leaving Jerusalem in the hands of the Turks."

"It seems as if God was against them, as he

was against the Jews when they undertook to rebuild their temple," said Ernest.

"Yes; the twelfth century seemed as much a 'leaden age' as the tenth. Men stood up for the truth, and were imprisoned and persecuted even unto death. You have read of the trials of the Waldenses."

"I have," said Mary, who, through all the Golden Hour conversations, had still kept her quiet place, even when she 'could not understand all;' "I have, in my 'Pierre and his Family.'"

"They were called Waldenses," resumed Mrs. Ewing, "because they were followers of one of these men who worked on the true building, in spite of the hosts of surrounding enemies. His name was Peter Waldo. He, too, like Bernard, was a Frenchman. God, through all these centuries of error, kept His Holy Word from destruction. And it was a powerful weapon in the hands of all who were attacked by false teachers. Waldo worked on the wall, with the weapon of God's word constantly at hand. From the four Gospels, which he translated into French, besides other portions of Scripture, he saw the errors into which those had fallen who 'departed from the simplicity of the gospel of Christ.' There was not nearly as much to be done as men supposed. Penances for accidental omissions and trifling mistakes were quite needless. He was a rich man, but, like the Master in whom he trusted, he distributed of his goods, and became poor, that others might through him have this word of life, and become rich. His translation was the first translation of Scripture into modern languages. The only version before this was the Latin Vulgate. Of course, he could not do so great a work as this without intense opposition. The enemies of the truth know the worth of the Bible, as well as its friends. This is why it has been chained, and burned, and forbidden to be used by the mass of the people. But God sets prisoners free in many ways, and it is as easy for Him to send an angel to unlock Bible chains as to open prison doors and iron gates for his people. Whenever the old cry is the cry of the worker still, 'Hear, O our God, for we are despised; and turn their reproach upon their own head, for they have provoked Thee to anger before the builders,' very soon the builders will be able to exclaim, 'So built we the wall.'"

"But they were persecuted," said Ernest.

"Yes. They were bitterly opposed, but their boldness for their faith did more for the advance of the work than if they had been left to go on undisturbed. Multitudes were led by their example of strong faith, and by their beautiful lives, to receive the same faith, and stand by it through any amount of opposition. Waldo himself, cursed by the pope, and driven from his home, took refuge in the mountains of Piedmont, where a great many followed him, sharing his cruel exile. Here was opened a grand quarry for the great temple. All Europe received more or less of the truth as it is in

Jesus through the open teachings and lives of these noble men. Communities were formed among these mountains, — pure as a flower amid Alpine snows. From these valleys 'the simple doctrine of Christianity flowed out in multiplied rivulets over all Europe, — Provence, Languedoc, Flanders, Germany, — until in the course of ages they swelled to a flood that swept over all lands.' When men think to 'hold the truth in unrighteousness,' God teaches them by such lessons as this that 'the word of God is not bound.' "

"Was Waldo a martyr, as well as an exile?" asked Olive. "I should think he would have been almost willing to be, when he saw how much good had been done by him."

"No doubt he would, had it been necessary; but exile was enough for him, and the witnessing of so much sorrow in those who shared his exile was probably more of a trial to bear than actual martyrdom. He traveled a great deal, teaching the doctrines he found in the Bible, and died at last, in Bohemia, in 1179. By his translations of Scripture he opened great windows of light in this building, and was thus like an agate; as it is said in Is. 54 : 12, 'And I will make thy windows of agate.' The people who had professed his views spread through many countries. They worked on in spite of persecution, which followed them as a sect from that day almost to this. It is scarcely a dozen years since they have been allowed full religious liberty in some of the countries where they are still found. Those who are in Italy have chosen Florence as their principal center, and have there a printing press, and a theological seminary, which was removed from Turin to Florence as recently as 1860. So you see what a work one man was enabled to start, and how, through his efforts, it has gone on, — God watching with the watchmen and working with the workers unto this day. For the light which entered at Waldo's windows of agate has never been shut out.

CHAPTER XIV.

CRYSTAL.

"WERE the crusaders satisfied with their defeats?" asked Ernest.

"No. In the thirteenth century they made several different attempts to gain Palestine, but with little success. It is supposed that the loss of life in these wars amounted to at least two millions."

"If they had only been true Christian martyrs!" said Olive.

"No doubt many among them were. They marched under the true banner, but, no doubt, with multitudes the symbol was very little regarded. Ambition, curiosity, and love of a wandering life induced many to enlist, no doubt, to whom the cross was a stumbling-block rather than a foundation of hope.

"In this century the Roman power was great, and seemed at times as if about to overshadow even the true faith of believers in Christ, as their Head. The pope took kings into hand, excommunicating them, and, taking them from their people, put other kings in their places. John, at that time king of England and Ireland, made all his kingdom formally over to the pope, binding himself by a public and formal document, in these words: 'I, John, by the grace of God king of England and lord of Ireland, for the expiation of my sins, and out of my own free will, with the advice and consent of my barons, do give unto the church of Rome and to Pope Innocent III. and his successors, the kingdoms of England and Ireland, together with all the rights belonging to them; and I will hold them of the pope as his vassal. I will be faithful to God, to the church of Rome, to the pope my lord, and to his successors lawfully appointed; and I bind myself to pay him a tribute of one thousand marks of silver yearly, viz., seven hundred for the kingdom of England, and three hundred for Ireland.' John was a brother of Richard of the Lion Heart, but not, like him, in favor of the spread of Christianity. He was a wicked, cruel king. This giving up of his kingdom to the pope brought great evil upon him. He died not long after, partly from mortification at being defeated in battle. But it has been said that his weakness wrought strength to the English nation. The Normans, who had, through him, become inhabitants of England, but were always heretofore under French rule, having through him lost their possessions, decided to regard the English as their countrymen, chose their rulers from among the English, and adopted England as their own country. So that, in the end, the wrath of man was made to be a praise, and the counsel of kings was brought to naught.

"There was more of superstition and error in the church in this century than in the dark days of the iron age. Many new orders of monks arose; useless rites and ceremonies increased; indulgences to sin were granted; the bread and wine became the soul and body of Christ, and rich caskets were made, in which bread, as God, might reside as in a dwelling, and be carried from place to place, which was probably the first idea of the elevation of the host.

"Among the festivals imposed upon the church in this century was the Year of Jubilee, which is still celebrated in Rome. It was said, first as a rumor, and afterward confirmed by the pope, that whoever should make a pilgrimage to St. Peter's Church, at Rome, in the years that finished centuries, — that is, within the last year or two of the close of the century, — they would merit by this indulgences for a hundred years. The pope sent word through all Christendom that whoever, at this particular time, should visit either St. Peter's or St. Paul's at Rome, confessing their sins, should be pardoned. It is a sad proof of the darkness of the time that such a proclamation could have produced any effect. The simple 'Come unto me' of Christ, had it been believed and obeyed, would have saved the long line of pilgrims making their weary and useless march from every part of Italy to its great capital in the year 1300. 'It has been estimated that two millions of people visited Rome in that year; and the concourse there was so great that many were trodden to death by the throng.'

"This was better success than the pope had looked for, and he and his people began to think one hundred years too long a time to pass between these pilgrimages. He might not live to see many such triumphs himself. So the next pope, Clement VI., shortened it half a century, and another line of pilgrims sought the pardon of their sins in 1350. Of course the next thing was to lessen the time again. Pope Nicholas V. settled the festival as one to be observed each quarter of a century."

"It makes me think of the officers' birthdays in the arctic winters," said Ernest.

"How was that?" asked Olive.

"Dr. Hayes said they had to devise every possible method of interesting the men in those cold, dark days, and among others, whenever an officer's birthday occurred, the cook took especial pains to get up an excellent dinner. He added, that it became apparent before long that some of the officers had more than their share of birthdays in the course of a year."

"During this century the Inquisition was established, the horrors of which I shall not attempt to tell you about. But through falsehood, fire, and sword, there were those who came out unscathed.

"It was little better in the fifteenth century. There was in this century a singular proof that the pope was not infallible, which, of course, his adherents failed to see; and that was, that two popes were to be found at the same time, — one in Rome, Urban VI., and the other at Avignon, Clement VII. It has never been clearly ascertained, as it was then altogether uncertain, which was the real, true pope. The worst of it was, that neither seemed to be proved immaculate, by the constant quarrels between themselves, which occasionally so jarred the whole building that many who still supposed them true, trembled for the foundations. While this perplexed the honest, conscientious people who believed the pope was the head of the church, it really strengthened the cause of Christians; for many, even kings, and the best men of the nation, saw easily through such presumption, and turned from a religion so grossly inconsistent and weak. One of the great builders of that century was John Wickliffe. You know he has been called the 'Morning Star of the Reformation.' They had long been working in darkness; and the work was severe and large, and they were separated upon the wall, one far from another. But the trumpet occasionally sounded a note of success, and some had held on until the stars began to appear. 'Our God shall fight for us,' might well have been their watchword as Wickliffe appeared. Afraid of neither popes, bishops, nor monks, of persecution or martyrdom, he boldly hewed, cut, and polished in the quarry of the Lord. He was born in Yorkshire, England, in or near 1324. He was educated at Oxford, a clear thinker, a sound scholar, and a keen detector of error. Theology was his favorite study. He wrote and preached against the prominent errors of the different orders of monks and friars. He became president of Baliol College, Oxford, and afterward warden of Canterbury, founded by an archbishop of that name. When the founder died, the right to his place was disputed, and after years of discussion, the pope and the king decided that he could not retain his position.

"Curiously enough, during this time he was called upon to pay his share of that same annual tribute of silver marks which King John had promised the pope. He refused to pay this tax, and wrote a paper declaring it useless, and having no warrant in reason or the Bible. Perhaps this was his first open opposition to the demands of the pope. He became a doctor of theology, and his views and teachings were soon highly regarded. He was at length charged openly with heresy, and had a partial trial before a synod called by the archbishop. But the people rose up in his favor, surrounded the building where the synod met, and rushed in to stand between him and harm. A royal messenger soon came, forbidding any severe measures, and with only some words of reprimand, and counsel not to repeat his offenses, he was allowed to retire.

"He was only the stronger for this opposition, and more fearless than before in maintaining his own views. Through him a complete translation of the Bible was given to the people. He had written copies multiplied, and in the hands of the people 'it became an engine of wonderful power.'

"I should think he was more than a pillar," said Olive.

"A pillar in a most important part of the building. This work of giving the Scriptures to others is a kind of building whose results are so great that eternity only can reveal them. Bede, and King Alfred, who translated the Psalms into Anglo-Saxon, did a great work; but Wickliffe's was the first complete translation, — which was published so as to be in general use, — from the Latin version, which, you remember, Jerome prepared, in the fourth century, from the original Hebrew and Greek."

"Was he a martyr?" asked Mary.

"No. God seemed to shield him from that,

for he was attacked, and attempts were made to try him for heresy; but he died at last peacefully, at the age of sixty, at his own rectory of Lutterworth. His doctrines were condemned after his death, and his enemies had the pleasure of showing whether they would have made him a martyr, by taking up his bones after they had lain quietly undisturbed for forty-one years, and publicly burning them, by order of the Council of Constance. The ashes were thrown into a little brook near by, called the Swift, which ran into the River Avon; and Thomas Fuller's remark, that their being thus scattered was emblematic of the spread of his doctrines, has been repeated in the stanza —

> 'The Avon to the Severn gave,
> The Severn to the sea;
> So Wickliffe's teachings shall be spread
> Wide as the waters be.' "

"Have we not workers in our time," asked Ernest, "something like Wickliffe? The missionaries, I mean, who have translated the Bible into the different languages of the heathen to whom they have gone."

"Yes; and we do not realize the greatness of the work which such noble men as Mills, and Hall, and Dwight, and Schauffler, and Judson, and Perkins, and Brown, and hosts of other wise builders, have done in this way. But God's eye is upon every worker, as well as upon every stone. He knows, better than the builders themselves, what riches of glory are yet in reserve for them, when, their last enemies having been subdued, and the fullness of time having come, the blessed promise shall be fulfilled —'Violence shall no more be heard in the land, wasting nor destruction within thy borders, but thou shalt call thy walls Salvation, and thy gates Praise.' Then those noble workers, who, like David, loved to walk about Zion and go round about her, telling the towers thereof, marking her bulwarks and considering her palaces, shall find that more of its beauty and safety came from their faithfulness than they had ever imagined while yet in the midst of the passing conflict."

"And we can help it a little by our pennies," said Mary.

"Nothing is lost, not even a cup of cold water given in the name of Christ. One verse from one torn leaf of a Bible has often been the ax and the hammer to fit a block for the beautiful temple. Wickliffe's life reminds us of crystal. He, too, served for a clear window, admitting the pure light of the Scriptures into this great temple."

CHAPTER XV.

COMPOSITION STONES.

"HAD never thought before," said Olive, "of the very many different beliefs, and forms of belief, regarding the way of salvation."

"And I have not named the half of them that were prevalent in these dark ages," replied Mrs. Ewing. "Some more absurd than any I have mentioned existed in Wickliffe's time. Persons who think dancing a part of worship, such as the Shakers in our time, may trace this to the fourteenth century. Men and women, holding each other by the hand, and dancing violently, professed to be favored with special revelations and spiritual visions. Another sect was more singular, but certainly as sincere, called the Flagellants. They believed that bodily punishment was as sure a way to obtain pardon as baptism, or any other easier method. And these persons went about in large crowds, through city streets and country towns, whipping themselves soundly as they went. They probably conceived the idea from the severity of the monks toward their own persons, and carried it to extremes, hoping for all the more full and speedy pardon. We see from it all how evil and bitter a thing sin is, and how much men will endure in the hope of ridding themselves of it. Perhaps the hardest lesson for the natural heart to learn is the ease with which the religion of Christ may be obtained.

"Notwithstanding the work of Wickliffe, great darkness prevailed throughout the fifteenth century. Constantinople was taken by the Turks in this century, and the Mohammedan religion, embraced by Turks and Tartars, spread throughout the East. The Greek church was almost destroyed, and Christians were bitterly opposed.

Many of the Greeks, however, leaving their own country, carried their learning and religion into the western countries of Europe, especially into Italy. In this century the art of printing became known, which, by rendering books cheaper, increased the light, and helped those who wished to spread the true knowledge of Christ.

"The popes of this age seemed far from perfect. Some of them were removed from their places by trials for the worst of crimes. Among them all, the most noted for wickedness was Roderic Borgia. Those who were no doubt truly pious in that age, — such men as Thomas à Kempis, for example, some of whose writings we still have, and parts of which are so purely devotional, — even those men remind us of composition stones, so made up were they in character of every variety of belief, mingled with superstitions. In this century, at one and the same time, they could boast of three popes; and their own quarrels, and the divisions of the church in consequence, worked great hindrances to the truth. This matter was settled by the

Council of Constance, in 1414. This council claimed to derive its power directly from Christ, and declared that whoever, even if the pope himself, refused obedience to their commands, should be punished. They were terribly severe in their punishments. Through them John Huss, a man of Wickliffe's spirit, bold in speaking and writing against the errors of the time, was condemned. He was a noble Christian, and he bore his sentence of death with a true martyr spirit. His enemies burned his works and destroyed his body; but his works still live, and to-day he rejoices in a martyr's crown. It was at this time that Wickliffe's bones were burned. Jerome also, a friend of Huss, more learned, and of great piety, was also condemned, and became a martyr soon after Huss, meeting his death in the same manner. The martyrdom of these two men, instead of discouraging, only stimulated the Christians to greater zeal. Wickliffe's followers in England, called Lollards, and the Waldenses, from their various hidden places of resort, still made themselves heard for the truth.

"It was in a time of darkness and persecution such as this that America was discovered. This country was graciously divided by the pope, Alexander VI., between the people of Spain and Portugal; and they were at once commanded to see that the people of this new world were taught in their religion. Monks of different orders were sent as missionaries, — the first to bring the news of the gospel in any form to our shores. There were great changes in the sixteenth century. While the popes still claimed to be Christ's true representatives, their lives were so corrupt that the people rose up against it in masses. They had the Scriptures in greater abundance, and some saw for themselves that such doctrines as that of indulgences for sin had no place in their teachings. These indulgences were sold to the people for money to build churches. Permits to commit sin, and absolution from all sin, 'how enormous soever,' were granted to those who would contribute to the building of St. Peter's Church at Rome. A corner-stone laid by indulgences to sin! A temple for worship built up by crime! Souls after death were to be released from purgatory by money. To the friends of such departed ones it was proclaimed that, 'as soon as the money tinkled in the chest, instant escape from torment and ascent to heaven' would follow.

"But there were men ready to say, as Peter said to Simon, 'Thy money perish with thee, because thou hast thought that the gift of God may be purchased with money!' One of these men rose up at this time from an unexpected quarter, who proved a wonderful builder, putting in windows of light, and opening long-barred doors, revealing to many countries the beauty of this temple. We will talk about him on the next Sabbath."

"I think I know who it was," said Ernest.

"Luther," said Olive.

"You are right," said Mrs. Ewing; "and we will call him the carbuncle, for you know, where it is said, 'Thy windows of agates,' it is added, and 'thy gates of carbuncle.'"

CHAPTER XVI.

CARBUNCLES.

"WHAT sort of stone is carbuncle?" asked Olive, the next Sabbath.

"It was a brilliant stone,—red, like glowing coals of fire, next in value to the diamond, and equal in hardness to the sapphire. Some writers have supposed it to be green in color, but it has generally been recognized as a flaming red. It was for its great hardness, and clearness, and beauty, that I chose it as a symbol of Luther, as well as because of the passage I quoted from Isaiah about the gates of carbuncle. If the gates of Zion are to be 'Praise,' then was Luther a fit symbol of such a part of the temple. He was a man of learning and great eloquence. His father, a poor miner, earned with his hard labor the money to educate his son, who early gave promise of future greatness. He intended him for a lawyer; God meant to make of him a wonderful builder on this great temple, struggling so slowly up through the centuries. In 1501, he entered the university at Erfurt. He was fond of reading, and after he had been over a year in this college, he one day found, in searching its library for something new, a copy of a Latin Bible. On reading it, strange feelings stirred within him. Not long after, he was seized with alarming illness. This led him to look upon life seriously. The sudden death of a friend by lightning added to these impressions, and leaving the college suddenly, he resolved on the life of a monk, and entered an Augustinian monastery. For years he continued in the church of Rome, but became a believer in the simple truths of the gospel by its study, and through the pious teachings of a friend."

"How did he break away from this life in the monastery?" asked Ernest.

"He was sent to Rome on some errand for the church, and seeing what corruption existed where he looked for perfection, he became fully convinced of their errors, and on his return preached with great boldness against the sins of the church, and its great errors of doctrine; showing the mighty contrast between their teachings and the simple truths of the Bible. 'The spark was laid to the train which ended in so mighty an explosion.' His boldness in proclaiming the truth aroused the fear of the cardinals, and of the pope himself, and he was excommunicated. For this he cared nothing. The thunder of the pope's bull had not half the power over him that the 'still small voice of the Spirit' had wrought in his soul. The pope then instructed the emperor to punish him as a rebel. The emperor, Charles V., would not do this without permitting him to speak in his own defense. A diet or council was called: at the city of Worms. Here Luther boldly set forth his belief, and made a most noble defense. He was allowed to go home without sentence of punishment, but on the way was secretly waylaid and imprisoned. He was declared by the emperor a heretic before this, but had a permission to return to his home safely, which was violated, and he was seized in a dense forest, and taken to the Castle of Wartburg, where, for a whole year, he was closely confined, his friends not knowing whether he was alive or dead. But here he was at work translating the New Testament into German, which was given to the people in 1522. For nine years, it is said, 'every year saw him publish a book, or books, against some form of papal error.' He translated much of the Bible, and wrote extensive commentaries upon it.

"Council after council met and 'decreed to suppress the Reformation by force;' but a greater decree than that of human councils had long before decided the matter, and they were again proving the truth of the verse I have so often quoted, 'Other foundation can no man lay.' Luther died in 1546. His last words were, 'Into Thy hands I commit my spirit; God of truth, Thou hast redeemed me.' He had lived to see Germany, in great part, set free from the papal power. There were many other noble workers of this time, friends of Luther. Melancthon was a friend and fellow-helper, and stood, perhaps, next to Luther in the great work of the Reformation. The name 'Protestant' was given at this time to those who *protested* against the doctrines of Rome. Luther's writings soon reached England, and were eagerly read by Wickliffe's followers.

"About this time, in 1524, a part of the New Testament was translated into English by William Tyndale. He has been called the noblest and greatest benefactor of the English race. He was educated at Oxford, and while Luther was opening a closed Bible for the Germans, he, without knowing it, was impelled to do the same for England."

"Was he then a Catholic?" asked Olive.

"Yes, and had been ordained a priest in 1502; but he put the truth before all titles, searched the Scriptures intently, and preached his convictions of truth boldly. In an earnest dispute with a learned papist divine, his zeal for the exact Scripture truth so irritated the man he was trying to convince, that he said to Tyndale, 'It were better for us to be without God's law than without the pope's.' To which Tyndale replied, 'I defy the pope and all his laws; and if God spare my life, ere many years I will cause a boy that driveth the plow to know more of the Scripture than you do.' It was a remark that meant something.

"He went to London to see about publishing the New Testament, with letters recommending him as a fine Greek scholar; but the bishop of London refused to aid or countenance him. In order to do this great work, he left England, and went to Cologne, where he began a translation; but before he had more than ten sheets struck off, his work was stopped. But he moved to another part of the wall, and began his work again at Worms, where Luther had made his defense. Three thousand copies were here soon finished in safety, and in the same year another edition was published. There is still shown, in the British Museum, one copy of this small octavo Bible. More than twenty editions followed this during the next ten years. Some of them were secretly conveyed into London in different ways. Once, two editions were sent, which were printed in Antwerp, conveyed in vessels loaded with grain."

"That was bread within bread," said Ernest.

"Yes. Once, when a number of these Antwerp Bibles remained on hand, — the bookseller who had them being a friend of Tyndale, — the bishop of Durham bought them all for the purpose of destroying them, and thought, when he had had them publicly burned, that he had made a great breach in the wall where

Tyndale was building. Instead of this, he was furnishing new material, for the money received for these Bibles was spent at once upon a new and better edition, which could not then have been published but for this very means. Tyndale made a pleasant use of this fact. When a prisoner of Sir Thomas More, his liberty was offered him if he would reveal the name of the man who aided him in defraying the expense of printing his Bibles. He replied that 'no one had done more than the bishop of Durham, for by paying a large sum for his books left on hand, he had enabled him to go on with good hope!' He was at length betrayed into the hands of his enemies, imprisoned at one time a year and a half, and, in 1536, sealed his faith by a martyr's death. But the tools he had prepared for other faithful workmen have been well used from that day to this. In the sixteenth century, also, Zuingle was doing the same work in Switzerland, and in France John Calvin was working like a second Luther,—all of them gates of praise to the temple."

"Was Calvin a martyr?" asked Mary.

"No; but he spent his life working boldly for the truth. Writing, preaching, visiting foreign countries, and spreading the knowledge of Bible truth in every possible way, he labored on, until, at the age of fifty-five, he, too, left the work to others."

"God always had a man ready," said Olive.

"In Scotland Knox was ready,—a man who adopted Luther's views, and boldly preached Bible doctrines without the fear of man. The enemies of the Christians were roused to attempt their utter destruction. In France, in 1572, an attack was made upon the Protestants, called the Massacre of St. Bartholomew. The palace bell tolled at Paris, as the signal for the dreadful work to commence, and for three days martyrs were multiplied by hundreds and thousands. More than thirty thousand men perished. This was joyful news to the pope, who ordered a jubilee to be held throughout all his dominions to celebrate the event.

"But for all this, the Christians not only continued to live among them, but to increase, and in 1598, under Henry IV., by the Edict of Nantes, the Protestants were granted equal rights with their enemies.

"In the seventeenth century the Jesuits visited China, as Japan had been entered the century before, and gained access to many in these countries, giving them some light and knowledge of the Scriptures, where now we are sending men to teach them the truth, as Luther, and Calvin, and Knox would have done. In England, during Luther's time, although he was opposed by Henry VIII., who himself wrote a book to refute Luther's teachings, yet this same king quarreled, from reasons of his own, with the pope, and broke his allegiance with him. Now, in the seventeenth century, great efforts were made to bring England back. The gunpowder plot, in 1605, was an attempt to blow up with powder the Parliament House, and thus kill the king and destroy the whole Parliament at once. The plot was discovered and defeated, and a similar attempt was afterward made, but without success. The wrath of popes, of emperors, and of kings has always spent itself in vain when set in opposition to God. He calls a Luther or a Wickliffe, a Whitefield or a Wesley, and shows how men can work when they work in a line with His will. Then none can hinder.

"From the time Christ said, 'Go ye into all the world, and preach the gospel to every creature,' His eye has been on each worker in every part of the field. There are builders to-day, cutters in marble, diggers for precious stones, blasting and hewing, sawing and polishing, so that if you listen you may hear the click of the hammer and the weaving of the saw in China and Japan, in Turkey and in Persia, in Africa and South America, and in the remote islands of the sea. Among the ruins of once great cities, His workmen are busy, where ancient towers and palaces, sand-buried and sea-buried, lie thick about them, and who are even now reminded, as they see old ruins torn down and robbed of their remaining pillars and arches, — some to be rebuilt into modern structures, and others to make way for the track of railroads, — that there is but one great temple that shall remain unchanged.

"What is it to these workers if the palace of Priam be destroyed for its building stones, or the temple of Ephesus leveled to the ground, if 'living stones' can be secured from these very waste places for an imperishable building? And, while fitting others for a place in this glorious temple, these self-denying workers are themselves becoming wonderful for strength, and beauty, and polish; and although they may not know it, and may feel, as Mary says, 'glad to be in anywhere,' God has in reserve, for every faithful worker, niches in this temple, where their beauty shall be seen when 'the gold for things of gold and silver for things of silver' shall have found their place, and when the 'onyx stones and stones to be set, glistering stones, and of divers colors, and all manner of precious stones,' shall not only have been prepared, but set in their own appropriate place in his living temple. The offerings of those who offered willingly shall not then be despised; for every willing, skillful man, for any manner of service, shall then have realized the promise, 'He will not fail thee nor forsake thee, until thou hast finished all the work for the service of the house of the Lord.'

"This promise nerves the arm of the workers to-day. Every foreign missionary, every home missionary, every president and officer of every mission board, every pastor, every Christian teacher, every Christian writer and Christian publisher, every benevolent Christian man, woman, or child, who is to-day working for Christ, in the strength of that promise, shall join in the shout when the top-stone is laid. And this shall be their cry, as they look upon the finished work of all the centuries, complete in its God-appointed beauty and glory: 'Thine, O Lord, is the greatness, and the power, and the glory, and the victory, and the majesty, for all that is in the heaven and in the earth is Thine. Thine is the kingdom, O Lord, and Thou art exalted above all.'"

"I like the 'Stone' title of Christ best of all," said Olive, as they closed their last Sunday afternoon's conversation.

"I should think every one would want to be a stone upon that foundation," said Mary.

"I have been thinking," said Ernest, "that some of the stones in this temple, such as Paul, and Jerome, and Wickliffe, and Luther, and Whitefield, are like loadstones, — they draw so many others to their true places in the building."

"Yes," said Mrs. Ewing, "it is a good thought. Christ is the great magnet. 'I will draw all men unto me,' he said."

"I do not see how any one can resist Him," said Olive.

"And yet there have always been those of whom it must be said, 'Oh, that they were wise! that they understood this!' Men build too much for the present, working for something that can be done at once, and seen of men. The great secret is to build for eternity; and this requires a foundation chosen from eternity, and which is eternal in its nature and duration. All who build upon Christ are safe for eternity, — safe as long as He is safe. Whether they stopped their work at early morning, and slept in a child's tiny casket, to waken amid the unseen glories of the beautiful city above; or whether they toiled till the stars of the morning came into view; or whether they stopped suddenly with their armor on in the heat of the day; whether their bodies were left to go down, unheard from, into the secret caverns of the sea, or to perish by fire or sword, or were disturbed by angry councils after death and given in ashes to swift rivers, or laid away, like Constantine's, in a coffin of gold, — they are all safe in the eye of Him who watches the dust of each one of His servants who has ever, since the world began, had a part in this wonderful building.

"In the day when He pronounces the work finished, they shall all be there. 'Every eye shall see Him,' and 'He will render to every man according to his work.' Then all who ever put their confidence in this precious corner stone, shall realize the blessedness of those who 'having not seen, yet believed.' And *so* shall the 'stone cut from the mountain without hands' FILL THE WHOLE EARTH."

THE END.

NEW PRIZE BOOKS.

I.

SHORT-COMINGS AND LONG-GOINGS. Price, $1.25. Full of sparkle and glow, and throbbing in every paragraph with intense life. It teaches the highest lessons of duty and religion with equal quietude and effect.

II.

LUTE FALCONER. Price, $1.50. A story of rare interest, touching the deeper chords of life. It will be read with enthusiasm, and laid down with an appreciation of its high office.

III.

THE JUDGE'S SONS. Price, $1.50. A well-told story, showing the worth of strong principles, and the success which follows pure and high aims. An admirable book to put into the hands of boys.

IV.

HESTER'S HAPPY SUMMER. It is rare to find a story of such sweetness and beauty. The pathos is tender and all-pervading, and steals into the heart with a refining power. The tone of the book is most healthy, making the world seem a better place to live in, because good people are found where least expected, and happiness is found, in any station, by unselfish lives. The influence of religion on different natures, and its ennobling power on all, are finely drawn.

V.

ONE YEAR OF MY LIFE. The form of a Diary is not attractive to lovers of story-books, but few can begin this Journal without reading to the end. It introduces to the minister's family, and makes one perfectly at home there, as if looking on and mingling with its daily life. The author is a finished writer, with a large knowledge of books and of life, a keen insight into character, a quick sense of the ludicrous, and a style of rare purity and ease.

VI.

BUILDING STONES. "A successful attempt to teach Bible truths in a manner both interesting and instructive. The curiosity is excited, and the mind kept on the alert, while important lessons are enforced. It is an admirable book for S. S. Libraries." — *Rev. Dr. Lincoln.*

OTHER VOLUMES IN PREPARATION.

Boston: D. Lothrop & Co., 38 and 40 Cornhill.

ANDY LUTTRELL. By Clara Vance. Price, $1.50. Andy Luttrell is an infidel — fierce, independent, almost desperate; a character of great power and no little genius, and handled so as to make a vivid and yet a healthful impression. This is a book for young men and young women. 16mo. ($500 prize series.) Illustrated.

GLENCOE PARSONAGE By Mrs. A. E. Porter. Price, $1.00.

STRAWBERRY HILL. By the author of "Andy Luttrell." Price, $1.50. The story especially exalts the settled faith and the working beneficence of the Gospel; making the possessor's life beautiful in itself, and the service redeeming to others. The characters are admirably drawn and developed. 16mo. Illustrated.

BLOOMFIELD. By Elizabeth Warren. Price, $1.50. Author of "The Story of Martin Luther," &c. An interesting and well-written English story, exhibiting piety in its unobtrusive aspects, as "steady, calm, meditative, and trustful," and which does not overlook home work in its devotion to that abroad, and insists upon a devout and loving spirit, even more than upon a bustling and restless activity. 16mo. Illustrated.

ANECDOTES OF ANIMALS; or, Two Great Families of the Animal Kingdom. Price, $1.50. By Mrs. Hugh Miller. With many beautiful illustrations.. 12mo.

OVERCOMING. By Mrs. E. K. Churchill. Price, $1.25. Etches the progress of a young person who is trying to follow conscience, against odds, to victory.

QUEEN OF THE ADRIATIC; or, Venice Past and Present. By W. H. D. Adams. Price, $1.75. With 31 illustrations. 12mo. 376 pp.

THE STORY OF POMPEII AND HERCULANEUM. This really valuable book gives a description of these ancient cities, and a compendium of the discoveries which have been made since 1713. Old and young can read this volume profitably, and enjoy it. 57 illustrations. Price, $1.25.

LITTLE BERTIE'S PICTURE LIBRARY. **Containing upwards of 100 illustrations. 12 Volumes, in Bright Bindings. Price, 25 cents each.**

LITTLE BLOSSOM STORIES. **5 Vols., 60 cents each. 119 illustrations.**

The Little Blossom Stories, and Bertie's Picture Library, are admirable. Compiled by Rev. J. D. STRONG, who knows what children want, and who brings to his task a true sympathy and a large skill, filled with charming and instructive stories, set off with an abundance of choice pictures, and bound in a style that honors beauty and good taste, we shall be surprised if there is not a steady and large demand for them wherever they are known.

SHINING HOURS. Price, $1.50. Willie, a noble, straight-forward, manly fellow, and his gentle, matter-of-fact sister Edith, being left orphans to the care of a plausible but dishonest guardian, who deceives and robs them, maintain a Christian spirit, continuing to be happy in spite of poverty and wrong, learning much wisdom and doing much good until prosperity returns to them. The story introduces a variety of characters, and leads through many adventures at home, at school, at college, and among the wild scenes of a hunter's life.

MASTER AND PUPIL. Price, $1.50. An admirably told story of school life. It goes straight to the heart, and makes religious character very lovely.

ARCHIBALD HAMILTON. Price, $1.25. A story of thrilling interest, teaching the grace of charity and virtue of forgiveness in most tender and forcible lessons.

JOHNNY JONES; or, the Bad Boy. Price, $1.00. A book of thrilling interest. All the boys will want it. 16mo. Illustrations by Billings.

BLUE-EYED JIMMY; or, the Good Boy. Price, $1.00. Illustrations by Billings.

THE RAINY DAY AT HOME. Price, 75 cents. With ten beautiful illustrations.

LITTLE BEN HADDEN; or, Do Right, whatever comes of it. By W. H. G. Kingston. A Story of the Sea. 16mo. Illustrated. Price, $1.25.

WONDERS OF CREATION. Price, 75 cents. Embracing Volcanoes and their phenomena. Twenty-five full-page illustrations, on tinted paper.

THE WONDERS OF NATURE. Price, 75 cents. Mammoth Cave, Dead Sea, High Mountains, Great Deserts, &c., with twenty-two illustrations.

WONDERS OF VEGETATION. Price, 75 cents. Describing the remarkable characteristics of some famous trees and singular plants. With 16 full-page illustrations.

WONDERS IN MANY LANDS. Price, 75 cents. Remarkable Rapids, Cascades, Waterfalls, Icebergs, Glaciers, Fountains, Natural Bridges, &c. Illustrated.

WILLIE MAITLAND; or, the Lord's Prayer Illustrated. Price, 55 cents.

WHEN WE WERE YOUNG. Price, 55 cents. By the Author of "A Trap to catch a Sunbeam." 18mo. Illustrated.

Choice Devotional and Gift Books.

QUIET HOUR LIBRARY. 6 Volumes. Price, 50 cents each. Per set, in a neat box, $3.00. Style of binding new and elegant. On tinted paper.

QUIET HOUR.	MORNING WATCHES.	NIGHT WATCHES.
MIND OF JESUS.	WORDS OF JESUS.	FAITHFUL PROMISER.

Very desirable as Presents in Sunday Schools, and as Christian gift books for all seasons.

ROCK OF AGES LIBRARY. Three elegant presentation volumes, quarto, beautifully printed on the finest plate paper, with red lines.

1. ROCK OF AGES. Poems, Selected and Original. By the Rev. S. F. SMITH, D.D., author of the hymn, "America," &c.
2. MORNING AND NIGHT WATCHES, AND THE QUIET HOUR.
3. MIND AND WORDS OF JESUS, AND FAITHFUL PROMISER.

In a handsome box. Price, $6.00.

DEVOTIONAL SERIES. 5 Volumes. Red Edges. Price, 75 cents per volume; full gilt, $1.25. Beautifully printed on fine paper, and in very elegant bindings.

MORNING AND NIGHT WATCHES. By Rev. J. R. MACDUFF.

THE RULE AND EXERCISES OF HOLY LIVING. By JEREMY TAYLOR, D.D.

MIND AND WORDS OF JESUS, AND FAITHFUL PROMISER. By Rev. J. R. MACDUFF.

IMITATION OF CHRIST. By THOS. A. KEMPIS.

THE RULE AND EXERCISES OF HOLY DYING. By JEREMY TAYLOR.

Make choice and appropriate presents for Pasto., Superintendent, or Teacher.

Boston: D. Lothrop & Co., 38 and 40 Cornhill.

GLENCOE PARSONAGE, Price, $1.00,

Is the journal of a clerg[illegible]'s wife, who, left a widow, has a son to bring up, whose tendencies and traits are a great trial to her; but after sad experience, and in answer to faith and prayer, all comes right at last. The book is full of interest

Boston: D. Lothrop & Co., 38 and 40 Cornhill.

HOURS OF CHRISTIAN DEVOTION. Translated from the German of Dr. A. Tholuck, with a preface by the Rev. Horatius Bonar, D.D.

"This new volume of Dr. Tholuck's will prove that there is no necessity for being flat and monotonous in a work of devotion, and that even in this region there can be found richness of thought and beauty of expression, that tell us we are in the society of a poet, and a thinker, and a critic, as well as of a man of God, and that makes us hesitate whether to designate these 'Hours of Devotion' a series of fine expositions, or an assemblage of lofty meditations, or a succession of noble hymns."

It is an elegant 16mo. volume, and will become one of the most popular gift books of the season Price, $2.00.

LEAVES OF COMFORT. Price, 75 cents. A Christian gift book for all seasons. 12mo., on tinted paper, with carmine border. This book contains all the poems which have been issued as leaflets, under the title of "Leaves of Comfort," and to which other choice poems have been added. It is issued in very tasty, attractive binding.

PILGRIM'S PROGRESS. Price, $3.50. An elegantly illustrated edition, on tinted paper, with handsome side and back dies, gilt edges, cloth, $3.50; Turkey morocco, $7.50; 32mo. edition, complete. Illustrated. 25 cents.

ROCK OF AGES. By Rev. S. F. SMITH, D.D. Author of the hymn, "America," &c. A choice collection of Religious Poems, original and selected. A new and elegantly printed gift book, bevelled boards, gilt edges, tinted paper, side die. Price, $1.50; Large edition, red lines, $2.00.

THE QUIET HOUR. Edited by the Rev. Heman Lincoln, D.D. Price, 50 cents. A book of prayer for every day in the week, and for special occasions.

LITTLE BEN HADDEN.

An excellent book, detailing the striking experiences of a lad who knew not a little both of hardship and temptation, but who, through it all, clung fast to the right, and won a noble victory. The volume gives much information respecting the various countries which the little hero visited during his life as a sailor.

Boston: D. Lothrop & Co., 38 and 40 Cornhill.

www.ingramcontent.com/pod-product-compliance
Lightning Source LLC
LaVergne TN
LVHW010244110826
845151LV00004B/1386

* 9 7 8 1 4 2 5 5 2 3 4 7 3 *